A Student's Guide to Vectors and Tensors

Vectors and tensors are among the most powerful problem-solving tools available, with applications ranging from mechanics and electromagnetics to general relativity. Understanding the nature and application of vectors and tensors is critically important to students of physics and engineering.

Adopting the same approach as in his highly popular *A Student's Guide to Maxwell's Equations*, Fleisch explains vectors and tensors in plain language. Written for undergraduate and beginning graduate students, the book provides a thorough grounding in vectors and vector calculus before transitioning through contra and covariant components to tensors and their applications. Matrices and their algebra are reviewed on the book's supporting website, which also features interactive solutions to every problem in the text, where students can work through a series of hints or choose to see the entire solution at once. Audio podcasts give students the opportunity to hear important concepts in the book explained by the author.

DANIEL FLEISCH is a Professor in the Department of Physics at Wittenberg University, where he specializes in electromagnetics and space physics. He is the author of *A Student's Guide to Maxwell's Equations* (Cambridge University Press, 2008).

A Student's Guide to Vectors and Tensors

DANIEL A. FLEISCH

CAMBRIDGE
UNIVERSITY PRESS

University Printing House, Cambridge CB2 8BS, United Kingdom

One Liberty Plaza, 20th Floor, New York, NY 10006, USA

477 Williamstown Road, Port Melbourne, VIC 3207, Australia

314-321, 3rd Floor, Plot 3, Splendor Forum, Jasola District Centre, New Delhi - 110025, India

79 Anson Road, #06-04/06, Singapore 079906

Cambridge University Press is part of the University of Cambridge.

It furthers the University's mission by disseminating knowledge in the pursuit of education, learning and research at the highest international levels of excellence.

www.cambridge.org
Information on this title: www.cambridge.org/9780521193696

First published 2012
9th printing 2018

A catalogue record for this publication is available from the British Library

ISBN 978-0-521-19369-6 Hardback
ISBN 978-0-521-17190-8 Paperback

Additional resources for this publication at www.cambridge.org/9780521171908

Contents

	Preface	*page* vii
	Acknowledgments	x
1	**Vectors**	1
1.1	Definitions (basic)	1
1.2	Cartesian unit vectors	5
1.3	Vector components	7
1.4	Vector addition and multiplication by a scalar	11
1.5	Non-Cartesian unit vectors	14
1.6	Basis vectors	20
1.7	Chapter 1 problems	23
2	**Vector operations**	25
2.1	Scalar product	25
2.2	Cross product	27
2.3	Triple scalar product	30
2.4	Triple vector product	32
2.5	Partial derivatives	35
2.6	Vectors as derivatives	41
2.7	Nabla – the del operator	43
2.8	Gradient	44
2.9	Divergence	46
2.10	Curl	50
2.11	Laplacian	54
2.12	Chapter 2 problems	60
3	**Vector applications**	62
3.1	Mass on an inclined plane	62
3.2	Curvilinear motion	72

3.3 The electric field 81
3.4 The magnetic field 89
3.5 Chapter 3 problems 95

4 Covariant and contravariant vector components 97
4.1 Coordinate-system transformations 97
4.2 Basis-vector transformations 105
4.3 Basis-vector vs. component transformations 109
4.4 Non-orthogonal coordinate systems 110
4.5 Dual basis vectors 113
4.6 Finding covariant and contravariant components 117
4.7 Index notation 122
4.8 Quantities that transform contravariantly 124
4.9 Quantities that transform covariantly 127
4.10 Chapter 4 problems 130

5 Higher-rank tensors 132
5.1 Definitions (advanced) 132
5.2 Covariant, contravariant, and mixed tensors 134
5.3 Tensor addition and subtraction 135
5.4 Tensor multiplication 137
5.5 Metric tensor 140
5.6 Index raising and lowering 147
5.7 Tensor derivatives and Christoffel symbols 148
5.8 Covariant differentiation 153
5.9 Vectors and one-forms 156
5.10 Chapter 5 problems 157

6 Tensor applications 159
6.1 The inertia tensor 159
6.2 The electromagnetic field tensor 171
6.3 The Riemann curvature tensor 183
6.4 Chapter 6 problems 192

Further reading 194
Index 195

Preface

This book has one purpose: to help you understand vectors and tensors so that you can use them to solve problems. If you're like most students, you first encountered vectors when you took a course dealing with mechanics in high school or college. At that level, you almost certainly learned that vectors are mathematical representations of quantities that have both magnitude and direction, such as velocity and force. You may also have learned how to add vectors graphically and by using their components in the x-, y- and z-directions.

That's a fine place to start, but it turns out that such treatments only scratch the surface of the power of vectors. You can harness that power and make it work for you if you're willing to delve a bit deeper – to see vectors not just as objects with magnitude and direction, but rather as objects that behave in very predictable ways when viewed from different reference frames. That's because vectors are a subset of a larger class of objects called "tensors," which most students encounter much later in their academic careers, and which have been called "the facts of the Universe." It is no exaggeration to say that our understanding of the fundamental structure of the universe was changed forever when Albert Einstein succeeded in expressing his theory of gravity in terms of tensors.

I believe, and I hope you'll agree, that tensors are far easier to understand if you first establish a stronger foundation in vectors, one that can help you cross the bridge between the "magnitude and direction" level and the "facts of the Universe" level. That's why the first three chapters of this book deal with vectors, the fourth chapter discusses coordinate transformations, and the last two chapters discuss higher-order tensors and some of their applications.

One reason you may find this book helpful is that if you spend a few hours looking through the published literature and on-line resources for vectors and tensors in physics and engineering, you're likely to come across statements such as these:

"A vector is a mathematical representation of a physical entity characterized by magnitude and direction."

"A vector is an ordered sequence of values."

"A vector is a mathematical object that transforms between coordinate systems in certain ways."

"A vector is a tensor of rank one."

"A vector is an operator that turns a one-form into a scalar."

You should understand that every one of these definitions is correct, but whether it's useful to you depends on the problem you're trying to solve. And being able to see the relationship between statements like these should prove very helpful when you begin an in-depth study of subjects that use advanced vector and tensor concepts. Those subjects include Mechanics, Electromagnetism, General Relativity, and others.

As with most projects, a good first step is to make sure you understand the terminology that will be used to attack the problem. For that reason, Chapter 1 provides the basic definitions you'll need to begin understanding vectors and tensors. And if you're ready for more-advanced definitions, you can find those at the beginning of Chapter 5.

You may be wondering how this book differs from other texts that deal with vectors and/or tensors. Perhaps the most important difference is that approximately equal weight is given to vector and tensor concepts, with one entire chapter (Chapter 3) devoted to selected vector applications and another chapter (Chapter 6) dedicated to example tensor applications.

You'll also find the presentation to be very different from that of other books. The explanations in this book are written in an informal style in which mathematical rigor is maintained only insofar as it doesn't obscure the underlying physics. If you feel you already have a good understanding of vectors and may need only a quick review, you should be able to skim through Chapters 1 through 3 very quickly. But if you're a bit unclear on some aspects of vectors and how to apply them to problems, you may find these early chapters quite helpful. And if you've already seen tensors but are unsure of exactly what they are or how to apply them, then Chapters 4 through 6 may provide some insight.

As a student's guide, this book comes with two additional resources designed to help you understand and apply vectors and tensors: an interactive website and a series of audio podcasts. On the website, you'll find the complete solution to every problem presented in the text in interactive format – that means you'll be able to view the entire solution at once, or ask for a series of helpful hints that will guide you to the final answer. So when you see a statement in the text saying that you can learn more about something by looking at the end-of-chapter problems, remember that the full solution to every one of those problems is available to you. And if you're the kind of learner who benefits from hearing spoken words rather than just reading text, the audio podcasts are for you. These MP3 files walk you through each chapter of the book, pointing out important details and providing further explanations of key concepts.

Is this book right for you? It is if you're a science or engineering student and have encountered vectors or tensors in one of your classes, but you're not confident in your ability to apply them. In that case, you should read the book, listen to the accompanying podcasts, and work through the examples and problems before taking additional classes or a standardized exam in which vectors or tensors may appear. Or perhaps you're a graduate student struggling to make the transition from undergraduate courses and textbooks to the more-advanced material you're seeing in graduate school – this book may help you make that step.

And if you're neither an undergraduate nor a graduate student, but a curious young person or a lifelong learner who wants to know more about vectors, tensors, or their applications in Mechanics, Electromagnetics, and General Relativity, welcome aboard. I commend your initiative, and I hope this book helps you in your journey.

Acknowledgments

It was a suggestion by Dr. John Fowler of Cambridge University Press that got this book out of the starting gate, and it was his patient guidance and unflagging support that pushed it across the finish line. I feel very privileged to have worked with John on this project and on my *Student's Guide to Maxwell's Equations*, and I acknowledge his many contributions to these books. A project like this really does take a village, and many others should be recognized for their efforts. While pursuing her doctorate in Physics at Notre Dame University, Laura Kinnaman took time to carefully read the entire manuscript and made major contributions to the discussion of the Inertia tensor in Chapter 6. Wittenberg graduate Joe Fritchman also read the manuscript and made helpful suggestions, as did Carnegie-Mellon undergraduate Wyatt Bridgeman. Carrie Miller provided the perspective of a Chemistry student, and her husband Jordan Miller generously shared his LaTeX expertise. Professor Adam Parker of Wittenberg University and Daniel Ross of the University of Wisconsin did their best to steer me onto a mathematically solid foundation, and Professor Mark Semon of Bates College has gone far beyond the role of reviewer and deserves credit for rooting out numerous errors and for providing several of the better explanations in this work. I alone bear the responsibility for any remaining inconsistencies or errors.

I also wish to acknowledge all the students who have taken a class from me during the two years it took me to write this book. I very much appreciate their willingness to share their claim on my time with this project. The greatest sacrifice has been made by the unfathomably understanding Jill Gianola, who gracefully accommodated the expanding time and space requirements of my writing.

1
Vectors

1.1 Definitions (basic)

There are many ways to define a vector. For starters, here's the most basic:

A vector is the mathematical representation of a physical entity that may be characterized by size (or "magnitude") and direction.

In keeping with this definition, speed (how fast an object is going) is not represented by a vector, but velocity (how fast and *in which direction* an object is going) does qualify as a vector quantity. Another example of a vector quantity is force, which describes how strongly and in what direction something is being pushed or pulled. But temperature, which has magnitude but no direction, is not a vector quantity.

The word "vector" comes from the Latin *vehere* meaning "to carry;" it was first used by eighteenth-century astronomers investigating the mechanism by which a planet is "carried" around the Sun.[1] In text, the vector nature of an object is often indicated by placing a small arrow over the variable representing the object (such as $\vec{F}$), or by using a bold font (such as $\boldsymbol{F}$), or by underlining (such as $\underline{F}$ or $\underset{\sim}{F}$). When you begin hand-writing equations involving vectors, it's very important that you get into the habit of denoting vectors using one of these techniques (or another one of your choosing). The important thing is not *how* you denote vectors, it's that you don't simply write them the same way you write non-vector quantities.

A vector is most commonly depicted graphically as a directed line segment or an arrow, as shown in Figure 1.1(a). And as you'll see later in this section, a vector may also be represented by an ordered set of N numbers,

[1] The *Oxford English Dictionary*. 2nd ed. 1989.

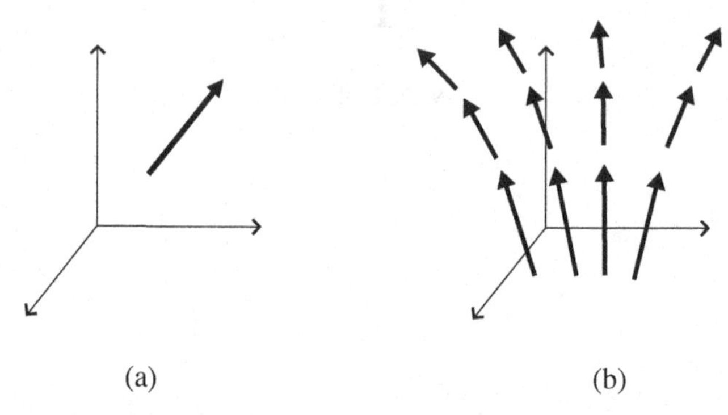

(a) (b)

Figure 1.1 Graphical depiction of a vector (a) and a vector field (b).

where N is the number of dimensions in the space in which the vector resides.

Of course, the true value of a vector comes from knowing what it represents. The vector in Figure 1.1(a), for example, may represent the velocity of the wind at some location, the acceleration of a rocket, the force on a football, or any of the thousands of vector quantities that you encounter in the world every day. Whatever else you may learn about vectors, you can be sure that every one of them has two things: size and direction. The magnitude of a vector is usually indicated by the length of the arrow, and it tells you the amount of the quantity represented by the vector. The scale is up to you (or whoever's drawing the vector), but once the scale has been established, all other vectors should be drawn to the same scale. Once you know that scale, you can determine the magnitude of any vector just by finding its length. The direction of the vector is usually given by indicating the angle between the arrow and one or more specified directions (usually the "coordinate axes"), and it tells you which way the vector is pointing.

So if vectors are characterized by their magnitude and direction, does that mean that two equally long vectors pointing in the same direction could in fact be considered to be the same vector? In other words, if you were to move the vector shown in Figure 1.1(a) to a different location without varying its length or its pointing direction, would it still be the same vector? In some applications, the answer is "yes," and those vectors are called free vectors. You can move a free vector anywhere you'd like as long as you don't change its length or direction, and it remains the same vector. But in many physics and engineering problems, you'll be dealing with vectors that apply *at a given location*; such vectors are called "bound" or "anchored" vectors, and you're not allowed to relocate bound vectors as you can free vectors.[2] You may see the term "sliding" vectors used for vectors that are free to move along their length but are not free to change length or direction; such vectors are useful for problems involving torque and angular motion.

You can understand the usefulness of bound vectors if you think about an application such as representing the velocity of the wind at various points in the atmosphere. To do that, you could choose to draw a bound vector at each point of interest, and each of those vectors would show the speed and direction of the wind at that location (most people draw the vector with its tail – the end without the arrow – at the point to which the vector is bound). A collection of such vectors is called a vector field; an example is shown in Figure 1.1(b).

If you think about the ways in which you might represent a bound vector, you may realize that the vector can be defined simply by specifying the start and end points of the arrow. So in a three-dimensional Cartesian coordinate system, you only need to know the values of x, y, and z for each end of the vector, as shown in Figure 1.2(a) (you can read about vector representation in non-Cartesian coordinate systems later in this chapter).

Now consider the special case in which the vector is anchored to the origin of the coordinate system (that is, the end without the arrowhead is at the point of intersection of the coordinate axes, as shown in Figure 1.2(b).[3] Such vectors may be completely specified simply by listing the three numbers that represent the x-, y-, and z-coordinates of the vector's end point. Hence a vector anchored to the origin and stretching five units along the x-axis may be represented as

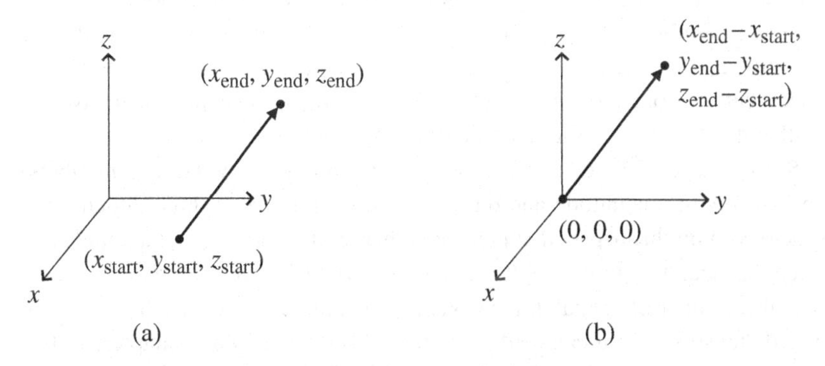

Figure 1.2 A vector in 3-D Cartesian coordinates.

[2] Mathematicians don't have much use for bound vectors, since the mathematical definition of a vector deals with how it transforms rather than where it's located.

[3] The vector shown in Figure 1.2 (a) can be shifted to this location by subtracting x_{start}, y_{start}, and z_{start} from the values at each end.

(5,0,0). In this representation, the values that represent the vector are called the "components" of the vector, and the number of components it takes to define a vector is equal to the number of dimensions in the space in which the vector exists. So in a two-dimensional space a vector may be represented by a pair of numbers, and in four-dimensional spacetime vectors may appear as lists of four numbers. This explains why a horizontal list of numbers is called a "row vector" and a vertical list of numbers is called a "column vector" in computer science. The number of values in such vectors tells you how many dimensions there are in the space in which the vector resides.

To understand how vectors are different from other entities, it may help to consider the nature of some things that are clearly *not* vectors. Think about the temperature in the room in which you're sitting – at each point in the room, the temperature has a value, which you can represent by a single number. That value may well be different from the value at other locations, but at any given point the temperature can be represented by a single number, the magnitude. Such magnitude-only quantities have been called "scalars" ever since W.R. Hamilton referred to them as "all values contained on the one scale of progression of numbers from negative to positive infinity."[4] Thus

A scalar is the mathematical representation of a physical entity that may be characterized by magnitude only.

Other examples of scalar quantities include mass, charge, energy, and speed (defined as the magnitude of the velocity vector). It is worth noting that the *change* in temperature over a region of space does have both magnitude and direction and may therefore be represented by a vector, so it's possible to produce vectors from groups of scalars. You can read about just such a vector (called the "gradient" of a scalar field) in Chapter 2.

Since scalars can be represented by magnitude only (single numbers) and vectors by magnitude and direction (three numbers in three-dimensional space), you might suspect that there are other entities involving magnitude and directions that are more complex than vectors (that is, requiring more numbers than the number of spatial dimensions). Indeed there are, and such entities are called "tensors."[5] You can read about tensors in the last three chapters of this book, but for now this simple definition will suffice:

[4] W.R. Hamilton, *Phil. Mag.* XXIX, 26.

[5] As you can learn in the later portions of this book, scalars and vectors also belong to the class of objects called tensors but have lower rank, so in this section the word "tensors" refers to higher-rank tensors.

A tensor is the mathematical representation of a physical entity that may be characterized by magnitude and multiple directions.

An example of a tensor is the inertia that relates the angular velocity of a rotating object to its angular momentum. Since the angular velocity vector has a direction and the angular momentum vector has a (potentially different) direction, the inertia tensor involves multiple directions.

And just as a scalar may be represented by a single number and a vector may be represented by a sequence of three numbers in 3-dimensional space, a tensor may be represented by an array of 3^R numbers in 3-dimensional space. In this expression, "R" represents the rank of the tensor. So in 3-dimensional space, a second-rank tensor is represented by $3^2 = 9$ numbers. In N-dimensional space, scalars still require only one number, vectors require N numbers, and tensors require N^R numbers.

Recognizing scalars, vectors, and tensors is easy once you realize that a scalar can be represented by a single number, a vector by an ordered set of numbers, and a tensor by an array of numbers. So in three-dimensional space, they look like this:

Scalar	Vector	Tensor (Rank 2)
(x)	(x_1, x_2, x_3) or $\begin{pmatrix} x_1 \\ x_2 \\ x_3 \end{pmatrix}$	$\begin{pmatrix} x_{11} & x_{12} & x_{13} \\ x_{21} & x_{22} & x_{23} \\ x_{31} & x_{32} & x_{33} \end{pmatrix}$

Note that scalars require no subscripts, vectors require a single subscript, and tensors require two or more subscripts – the tensor shown here is a tensor of rank 2, but you may also encounter higher-rank tensors, as discussed in Chapter 5. A tensor of rank 3 may be represented by a three-dimensional array of values.

With these basic definitions in hand, you're ready to begin considering the ways in which vectors can be put to use. Among the most useful of all vectors are the Cartesian unit vectors, which you can read about in the next section.

1.2 Cartesian unit vectors

If you hope to use vectors to solve problems, it's essential that you learn how to handle situations involving more than one vector. The first step in that process is to understand the meaning of special vectors called "unit vectors" that often

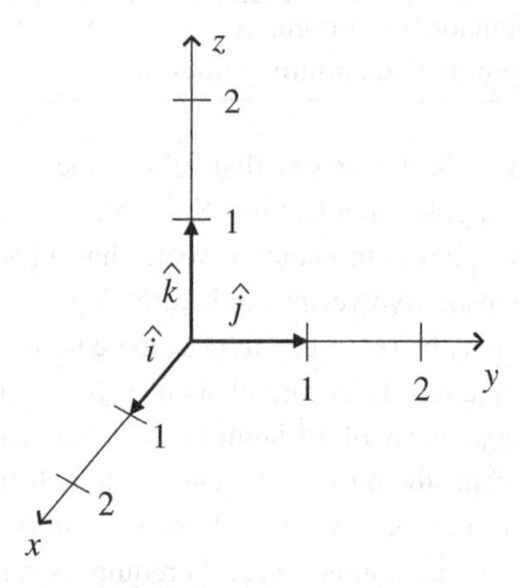

Figure 1.3 Unit vectors in 3-D Cartesian coordinates.

serve as markers for various directions of interest (unit vectors may also be called "versors").

The first unit vectors you're likely to encounter are the unit vectors $\hat{x}$, $\hat{y}$, $\hat{z}$ (also called $\hat{\imath}$, $\hat{\jmath}$, $\hat{k}$) that point in the direction of the x-, y-, and z-axes of the three-dimensional Cartesian coordinate system, as shown in Figure 1.3. These vectors are called unit vectors because their length (or magnitude) is always exactly equal to unity, which is another name for "one." One what? One of whatever units you're using for that axis.

You should note that the Cartesian unit vectors $\hat{\imath}$, $\hat{\jmath}$, $\hat{k}$ can be drawn at any location, not just at the origin of the coordinate system. This is illustrated in Figure 1.4. As long as you draw a vector of unit length pointing in the same direction as the direction of the (increasing) x-axis, you've drawn the $\hat{\imath}$ unit vector. So the Cartesian unit vectors show you the directions of the x, y, and z axes, *not* the location of the origin.

As you'll see in Chapter 2, unit vectors can be extremely helpful when doing certain operations such as specifying the portion of a given vector pointing in a certain direction. That's because unit vectors don't have their own magnitude to throw into the mix (actually, they do have their own magnitude, but it is always one).

So when you see an expression such as "$5\hat{\imath}$," you should think "5 units along the positive x-direction." Likewise, $-3\hat{\jmath}$ refers to 3 units along the negative y-direction, and $\hat{k}$ indicates one unit along the positive z-direction.

Of course, there are other coordinate systems in addition to the three perpendicular axes of the Cartesian system, and unit vectors exist in those coordinate

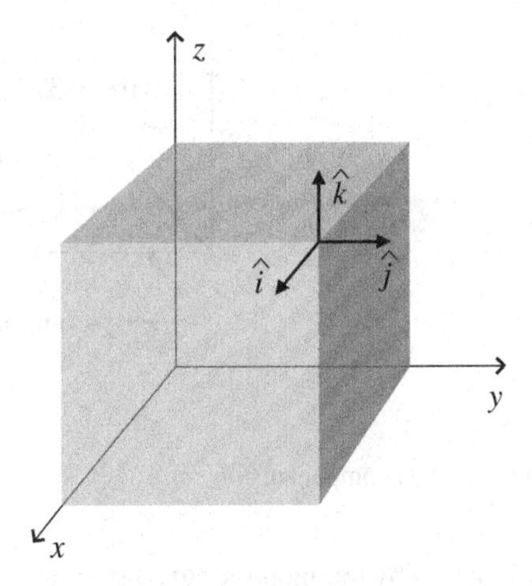

Figure 1.4 Cartesian unit vectors at an arbitrary point.

systems as well; you can see some examples in Section 1.5. One advantage of the Cartesian unit vectors is that they point in the same direction no matter where you go; the x-, y-, and z-axes run in straight lines all the way out to infinity, and the Cartesian unit vectors are parallel to the directions of those lines everywhere.

To put unit vectors such as $\hat{\imath}$, $\hat{\jmath}$, $\hat{k}$ to work, you need to understand the concept of vector components. The next section shows you how to represent vectors using unit vectors and vector components.

1.3 Vector components

The unit vectors described in the previous section are especially useful when they become part of the "components" of a vector. And what are the components of a vector? Simply stated, they are the pieces that can be used to make up the vector.

To understand vector components, think about the vector $\vec{A}$ shown in Figure 1.5. This is a bound vector, anchored at the origin and extending to the point ($x = 0$, $y = 3$, $z = 3$) in a three-dimensional Cartesian coordinate system. So if you consider the coordinate axes as representing the corner of a room, this vector is embedded in the back wall (the yz plane).

Imagine you're trying to get from the beginning of vector $\vec{A}$ to the end – the direct route would be simply to move in the direction of the vector. But if you were constrained to move only in the directions of the axes, you could get from

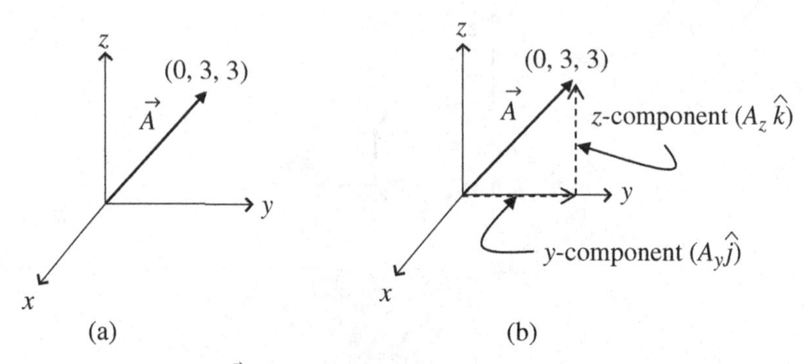

Figure 1.5 Vector $\vec{A}$ and its components.

the origin to your destination by taking three (unit) steps along the y-axis, then turning 90° to your left, and then taking three more (unit) steps in the direction of the z-axis.

What does this little journey have to do with the components of a vector? Simply this: the lengths of the components of vector $\vec{A}$ are the distances you traveled in the directions of the axes. Specifically, in this case the magnitude of the y-component of vector $\vec{A}$ (written as A_y) is just the distance you traveled in the direction of the y-axis (3 units), and the magnitude of the z-component of vector $\vec{A}$ (written as A_z) is the distance you traveled in the direction of the z-axis (also 3 units). Since you didn't move at all in the direction of the x-axis, the magnitude of the x-component of vector $\vec{A}$ (written as A_x) is zero.

A very handy and compact way of writing a vector as a combination of vector components is this:

$$\vec{A} = A_x\hat{\imath} + A_y\hat{\jmath} + A_z\hat{k}, \tag{1.1}$$

where the magnitudes of the vector components (A_x, A_y, and A_z) tell you how many unit steps to take in each direction ($\hat{\imath}, \hat{\jmath}$, and $\hat{k}$) to get from the beginning to the end of vector $\vec{A}$.[6]

When you read about vectors and vector components, you're likely to run across statements such as "The components of a vector are the projections of the vector onto the coordinate axes." As you can see in Chapter 4, exactly how those projections are made can have a significant influence on the nature of the components you get. But in Cartesian coordinate systems (and other

[6] Some authors refer to the magnitudes A_x, A_y, and A_z as the "components of $\vec{A}$," while others consider the components to be $A_x\hat{\imath}$, $A_y\hat{\jmath}$, and $A_z\hat{k}$. Just remember that A_x, A_y, and A_z are scalars, but $A_x\hat{\imath}$, $A_y\hat{\jmath}$, and $A_z\hat{k}$ are vectors.

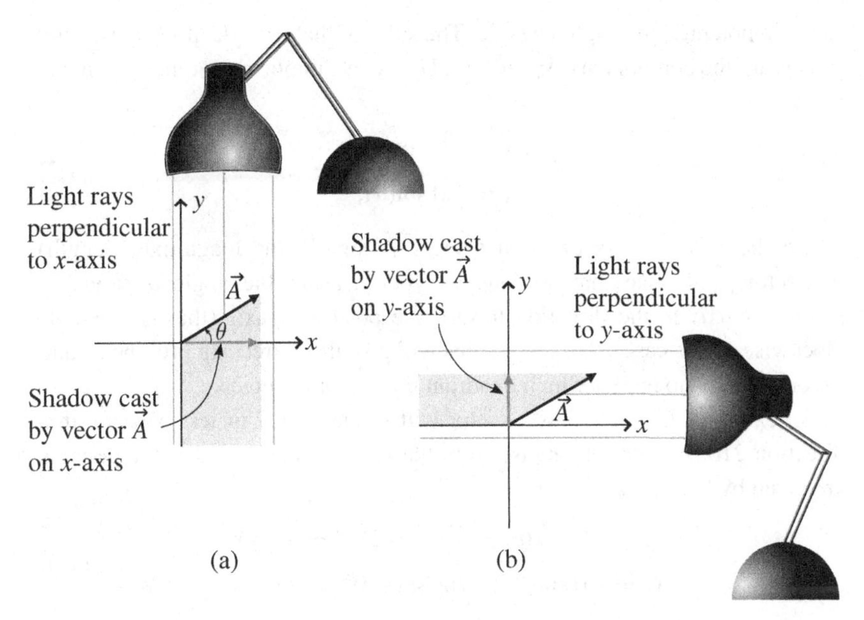

Figure 1.6 Vector components as projections onto x- and y-axes.

"orthogonal" systems in which the axes are perpendicular to one another), the concept of projection onto the coordinate axes is unambiguous and may be very helpful in picturing the components of a vector.

To understand how this works, take a look at vector $\vec{A}$ and the light sources and shadows in Figure 1.6. As you can see in Figure 1.6(a), the direction of the light that produces the shadow on the x-axis is parallel to the y-axis (actually antiparallel since it's moving in the negative y-direction), which in this case is the same as saying that the direction of the light is perpendicular to the x-axis.

Likewise, in Figure 1.6(b), the direction of the light that produces the shadow on the y-axis is antiparallel to the x-axis, which is of course perpendicular to the y-axis. This may seem like a trivial point, but when you encounter non-orthogonal coordinate systems, you'll find that the direction parallel to one axis is not necessarily perpendicular to another axis, which gives rise to an entirely different type of vector component. This simple fact has profound implications for the behavior of vectors and tensors for observers in different reference frames, as you'll see in Chapters 4, 5, and 6.

No such issues arise in the two-dimensional Cartesian coordinate system shown in Figure 1.6, and in this case the magnitudes of the components of vector $\vec{A}$ are easy to determine. If the angle between vector $\vec{A}$ and the positive x-axis is θ, as shown in Figure 1.6a, it's clear that the length of $\vec{A}$ can be seen as the hypotenuse of a right triangle. The sides of that triangle along the x- and y-axes are the components A_x and A_y. Hence by simple trigonometry you can write:

$$\begin{aligned} A_x &= |\vec{A}|\cos(\theta), \\ A_y &= |\vec{A}|\sin(\theta), \end{aligned} \tag{1.2}$$

where the vertical bars on each side of $\vec{A}$ signify the magnitude (length) of vector $\vec{A}$. Notice that so long as you measure the angle θ *from the positive x-axis* in the direction toward the positive y-axis (that is, counter-clockwise in this case), these equations will give the correct sign for the x- and y-components no matter which quadrant the vector occupies.

For example, if vector $\vec{A}$ is a vector with a length of 7 meters pointing in a direction 210° counter-clockwise from the $+x$-axis, the x- and y-components are given by Eq. 1.2 as

$$\begin{aligned} A_x &= |\vec{A}|\cos(\theta) = 7\text{m}\ \cos 210^\circ = -6.1\,\text{m}, \\ A_y &= |\vec{A}|\sin(\theta) = 7\text{m}\ \sin 210^\circ = -3.5\,\text{m}. \end{aligned} \tag{1.3}$$

As expected for a vector pointing down and to the left from the origin, both components are negative.

It's equally straightforward to find the length and direction of a vector if you're given the vector's Cartesian components. Since the vector forms the hypotenuse of a right triangle with sides A_x and A_y, the Pythagorean theorem tells you that the length of $\vec{A}$ must be

$$|\vec{A}| = \sqrt{A_x^2 + A_y^2}, \tag{1.4}$$

and from trigonometry

$$\theta = \arctan\left(\frac{A_y}{A_x}\right), \tag{1.5}$$

where θ is measured counter-clockwise from the positive x-axis in a right-handed coordinate system. If you try this with the components of vector $\vec{A}$ from Eq. 1.3 and end up with a direction of 30° rather than 210°, remember that unless you have a four-quadrant arctan function on your calculator, you must add 180° to the angle whenever the denominator of the expression (A_x in this case) is negative.

Once you have a working understanding of unit vectors and vector components, you're ready to do basic vector operations. The entirety of Chapter 2 is devoted to such operations, but two of them are needed for the remainder of this chapter. For that reason, you can read about vector addition and multiplication by a scalar in the next section.

1.4 Vector addition and multiplication by a scalar

If you've read the previous section on vector components, you've already seen two vector operations in action. Those two operations are the addition of vectors and multiplication of a vector by a scalar. Both of these operations are used in the expansion of a vector in terms of vector components as in Eq. 1.1 from Section 1.3:

$$\vec{A} = A_x\hat{\imath} + A_y\hat{\jmath} + A_z\hat{k}.$$

In each of these terms, the unit vector ($\hat{\imath}$, $\hat{\jmath}$, or $\hat{k}$) is being multiplied by a scalar (A_x, A_y, or A_z), and you already know the effect of that: it produces a new vector, in the same direction as the unit vector, but longer than unity by the value of the component (or shorter if the magnitude of the component is between zero and one). So multiplying a vector by any positive scalar does not change the direction of the vector, but only scales the length of the vector. Hence, $5\vec{A}$ is a vector in exactly the same direction as $\vec{A}$, but with length five times that of $\vec{A}$, as shown in Figure 1.7(a). Likewise, multiplying $\vec{A}$ by (1/2) produces a vector that points in the same direction as $\vec{A}$ but is only half as long. So the vector component $A_x\hat{\imath}$ is a vector in the $\hat{\imath}$ direction, but with length A_x units (since $\hat{\imath}$ has a length of one unit).

There is a caveat that goes with the "changes length, not direction" rule when multiplying a vector by a scalar: if the scalar is *negative*, then the vector is reversed in direction in addition to being scaled in length. Thus multiplying vector $\vec{B}$ by -2 produces the new vector $-2\vec{B}$, and that vector is twice as long as $\vec{B}$ and points in the opposite direction to $\vec{B}$, as shown in Figure 1.7(b).

The other operation going on in Eq. 1.1 is vector addition, and you already have an idea of what that means if you recall Figure 1.5 and the process of getting from the beginning of vector $\vec{A}$ to the end. In that process, the quantity

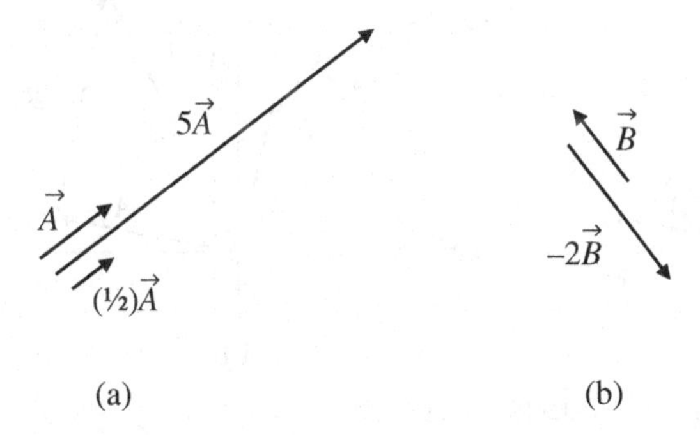

Figure 1.7 Multiplication of a vector by a scalar.

$A_y\hat{\jmath}$ represented not only the number of steps you took, but also the direction in which you took them. Likewise, the quantity $A_z\hat{k}$ represented the number of steps you took *in a different direction*. The fact that these two quantities include directional information means that you cannot simply add them together algebraically; you must add them "as vectors."

To accomplish vector addition graphically, you simply imagine moving one vector (without changing its length or direction) so that its tail is at the head of the other vector. The sum is then determined by making a new vector that begins at the start of the first vector and terminates at the end of the second vector. You can do this graphically, as in Figure 1.5(b), where the tail of vector $A_z\hat{k}$ is placed at the head of vector $A_y\hat{\jmath}$, and the sum is the vector from the beginning of $A_y\hat{\jmath}$ to the end of $A_z\hat{k}$.

This graphical "head-to-tail" approach to vector addition works for any vectors (and any number of vectors), not just two vectors that are perpendicular to one another (as $A_y\hat{\jmath}$ and $A_z\hat{k}$ were). An example of this is shown in Figure 1.8. To graphically add the two vectors $\vec{A}$ and $\vec{B}$ in Figure 1.8(a), you simply imagine moving one of the two vectors so that its tail is at the position of the other vector's head (it doesn't matter which vector you choose to move; the result will be the same). This is illustrated in Figure 1.8(b), in which vector $\vec{B}$ has been displaced so that its tail is at the head of vector $\vec{A}$. The sum of these two vectors (called the "resultant" vector $\vec{C} = \vec{A} + \vec{B}$) is the vector that extends from the beginning of $\vec{A}$ to the end of $\vec{B}$.

Knowing how to add vectors graphically means you can always determine the sum of two or more vectors simply using a ruler and a protractor; just draw the vectors head-to-tail (being careful to maintain each vector's length and

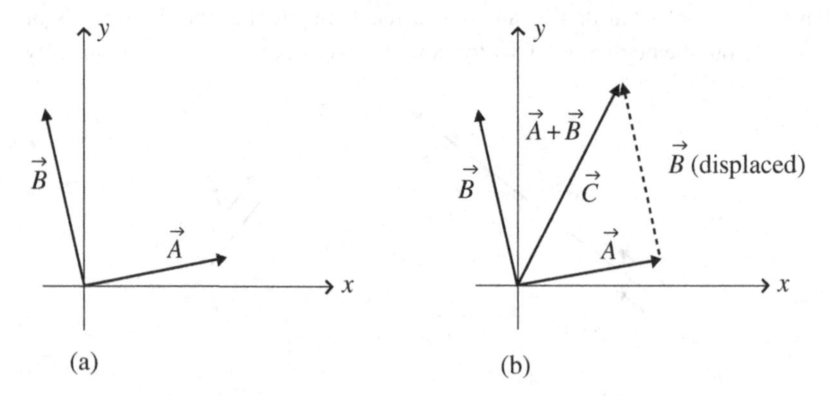

Figure 1.8 Graphical addition of vectors.

angle), sketch the resultant from the beginning of the first to the end of the last, and then measure the length (using the ruler) and angle (using the protractor) of the resultant. This approach can be both tedious and inaccurate, so here's an alternative approach that uses the components of each vector: if vector $\vec{C}$ is the sum of two vectors $\vec{A}$ and $\vec{B}$, then the magnitude of the x-component of vector $\vec{C}$ (which is just C_x) is the sum of the magnitudes of the x-components of vectors $\vec{A}$ and $\vec{B}$ (that is, $A_x + B_x$), and the magnitude of the y-component of vector $\vec{C}$ (called C_y) is the sum of the magnitudes of the y-components of vectors $\vec{A}$ and $\vec{B}$ (that is, $A_y + B_y$). Thus

$$\begin{aligned} C_x &= A_x + B_x, \\ C_y &= A_y + B_y. \end{aligned} \tag{1.6}$$

The rationale for this is shown in Figure 1.9.

Once you have the components C_x and C_y of the resultant vector $\vec{C}$, you can find the magnitude and direction of $\vec{C}$ using

$$|\vec{C}| = \sqrt{C_x^2 + C_y^2} \tag{1.7}$$

and

$$\theta = \arctan\left(\frac{C_y}{C_x}\right) \tag{1.8}$$

To see how this works in practice, imagine that vector $\vec{A}$ in Figure 1.9 is given by $\vec{A} = 6\hat{\imath} + \hat{\jmath}$ and vector $\vec{B}$ is given by $\vec{B} = -2\hat{\imath} + 8\hat{\jmath}$. To add these two vectors algebraically, you simply use Eqs. 1.6:

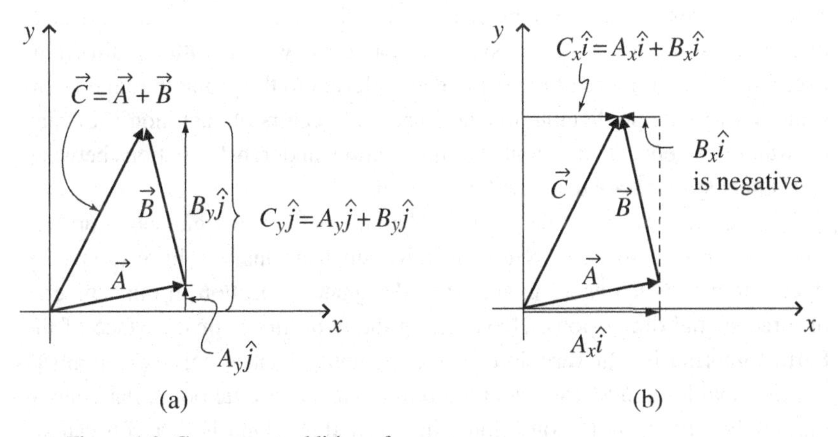

Figure 1.9 Component addition of vectors.

$$C_x = A_x + B_x = 6 + (-2) = 4,$$
$$C_y = A_y + B_y = 1 + 8 = 9,$$

so $\vec{C} = 4\hat{\imath} + 9\hat{\jmath}$. If you wish to know the magnitude of $\vec{C}$, you can just plug the components into Eq. 1.7 to get

$$|\vec{C}| = \sqrt{C_x^2 + C_y^2} = \sqrt{4^2 + 9^2}$$
$$= \sqrt{16 + 81} = 9.85.$$

And the angle that $\vec{C}$ makes with the positive x-axis is given by Eq. 1.8:

$$\theta = \arctan\left(\frac{C_y}{C_x}\right)$$
$$= \arctan\left(\frac{9}{4}\right) = 66.0°.$$

With the basic operations of vector addition and multiplication of a vector by a scalar in hand, you're ready to begin thinking about the more advanced uses of vectors. But you're also ready to attack a variety of problems involving vectors, and you can find a set of such problems at the end of this chapter.[7]

1.5 Non-Cartesian unit vectors

The three straight, mutually perpendicular axes of the Cartesian coordinate system are immensely useful for a variety of problems in physics and engineering. Some problems, however, are much easier to solve in other coordinate systems, often because the axes of those systems more closely align with the directions over which one or more of the parameters relevant to the problem remain constant or vary in a predictable manner. The unit vectors of such non-Cartesian coordinate systems are the subject of this section, and transformations between coordinate systems are discussed in Chapter 4.

As described earlier, it takes exactly N numbers to unambiguously represent any location in a space of N dimensions, which means you have to specify three numbers (such as x, y, and z) to designate a location in our Universe of three spatial dimensions. However, on the two-dimensional surface of the Earth (ignoring height variation for the moment) it takes only two numbers (latitude and longitude, for example) to designate a specific point. And one of the few benefits to living on a long, infinitely thin island is that you can set

[7] Remember that full solutions are available on the book's website.

up a rendezvous using only a single number to describe the location ("I'll be waiting for you at 3.75 kilometers").

Of course, numbers define locations only after you've defined the *coordinate system* that you're using. For example, do you mean 3.75 kilometers from the east end of the island or from the west end? In every space of 1, 2, 3, or more dimensions, you can devise an infinite number of coordinate systems to specify locations in that space. In each of those coordinate systems, at each location there's one direction in which one of the coordinates is increasing the fastest, and if you lay a vector with length of one unit in that direction, you've defined a coordinate unit vector for that system. So in the Cartesian coordinate system, the $\hat{\imath}$ unit vector shows you the direction in which the x-coordinate increases, the $\hat{\jmath}$ unit vector shows you the direction in which the y-coordinate increases, and the $\hat{k}$ unit vector shows you the direction in which the z-coordinate increases. Other coordinate systems have their own coordinate unit vectors, as well.

Consider the two-dimensional coordinate systems shown in Figure 1.10. In a two-dimensional space, you know that it takes two numbers to specify any location, and those numbers could be x and y, defined along two straight axes that intersect at a right angle. The x value tells you how far you are to the right of the y-axis (or to the left if the x value is negative), and the y value tells you how far you are above the x-axis (or below if the y value is negative). But you could equally well specify any location in this two-dimensional space by noting how far and in what direction you've moved from the origin. In the standard version of these "polar" coordinates, the distance from the origin is

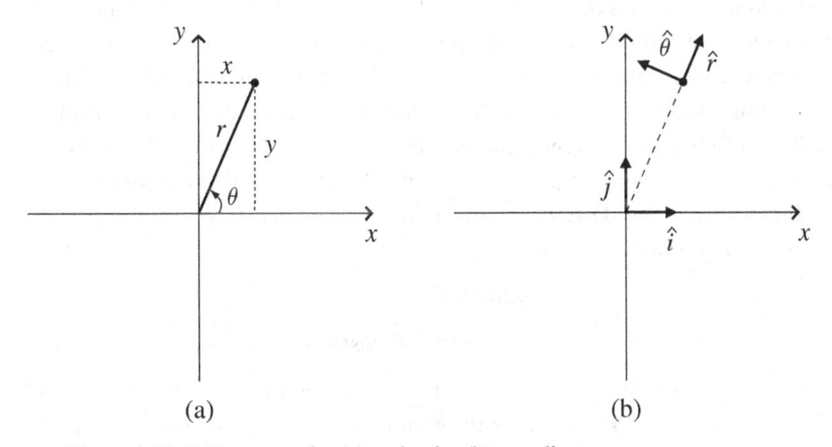

Figure 1.10 2-D rectangular (a) and polar (b) coordinates.

called r and the direction is specified by giving the angle θ measured counter-clockwise from the positive x-axis.

It's easy enough to figure out one set of coordinates if you know the others; for example, if you know the values of x and y, you can find r and θ using

$$\begin{aligned} r &= \sqrt{x^2 + y^2} \\ \theta &= \arctan\left(\frac{y}{x}\right). \end{aligned} \tag{1.9}$$

Likewise, if you have the values of r and θ, you can find x and y using

$$\begin{aligned} x &= r\cos(\theta) \\ y &= r\sin(\theta). \end{aligned} \tag{1.10}$$

For the point shown in Figure 1.10, if the values of x and y are 4 cm and 9 cm, then r has a value of approximately 9.85 cm and θ has a value of 66.0°. Clearly, whether you write $(x, y) = (4\text{ cm}, 9\text{ cm})$ or $(r, \theta) = (9.85\text{ cm}, 66.0°)$, you're referring to the same location; it's not the point that's changed, it's only the point's coordinates that are different.

And if you choose to use the polar coordinate system to represent the point, do unit vectors exist that serve the same function as $\hat{\imath}$ and $\hat{\jmath}$ in Cartesian coordinates? They certainly do, and with a little logic you can figure out which direction they must point. After all, you know that the unit vector $\hat{\imath}$ shows you the direction of increasing x and the unit vector $\hat{\jmath}$ shows you the direction of increasing y, but now you're using r and θ instead of x and y. So it seems reasonable that the unit vector $\hat{r}$ at any location should point in the direction of increasing r, and the unit vector $\hat{\theta}$ should point in the direction of increasing θ. For the point shown in Figure 1.10, that means that $\hat{r}$ should point up and to the right, in the direction of increasing r if θ is held constant. At that same point, $\hat{\theta}$ should point up and to the left, in the direction of increasing θ if r is held constant. These polar unit vectors are shown for one point in Figure 1.10(b).

An important consequence of this definition is that the directions of $\hat{r}$ and $\hat{\theta}$ will be different at different locations. They'll always be perpendicular to one another, but they will not point in the same directions as they do for the point in Figure 1.10. The dependence of the polar unit vectors on position can be seen in the following relations:

$$\begin{aligned} \hat{r} &= \cos(\theta)\hat{\imath} + \sin(\theta)\hat{\jmath} \\ \hat{\theta} &= -\sin(\theta)\hat{\imath} + \cos(\theta)\hat{\jmath}. \end{aligned} \tag{1.11}$$

So if $\theta = 0$ (which means your location is on the $+x$-axis), then $\hat{r} = \hat{\imath}$ and $\hat{\theta} = \hat{\jmath}$. But if $\theta = 90°$ (so your location is on the $+y$-axis), then $\hat{r} = \hat{\jmath}$ and $\hat{\theta} = -\hat{\imath}$.

Does this dependence on position mean that these unit vectors are not "real" vectors? That depends on your definition of a real vector. If you define a vector as a quantity with magnitude and direction, the polar unit vectors do meet your definition. But they do not meet the definition of free vectors described in Section 1.1, since they may not be moved without changing their direction.

This means that if you express a vector in polar coordinates and then take the derivative of that vector, you'll have to account for the change in the unit vectors, as well. That's one of the advantages offered by Cartesian coordinates – the unit vectors do not change no matter where you go in the space.

As you might expect, the situation is slightly more complicated for three-dimensional coordinate systems. Whether you choose to use Cartesian or non-Cartesian coordinates, you're going to need three variables to represent all the possible locations in a three-dimensional space, and each of the coordinates is going to come with its own unit vector. The two most common three-dimensional non-Cartesian coordinate systems are cylindrical and spherical coordinates, which you can see in Figures 1.11 and 1.12.

In cylindrical coordinates a point P is specified by r, ϕ, z, where r (sometimes called ρ) is the perpendicular distance from the z-axis, ϕ is the angle measured from the x-axis to the projection of r onto the xy plane, and z is the same as the z in Cartesian coordinates. Here's how you find r, ϕ, and z if you know x, y, and z:

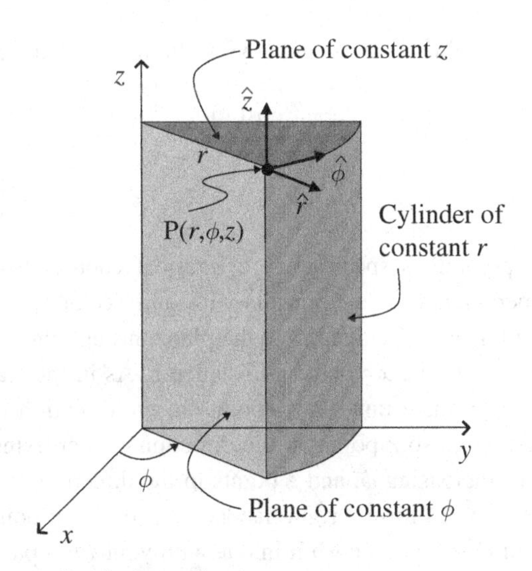

Figure 1.11 Cylindrical coordinates.

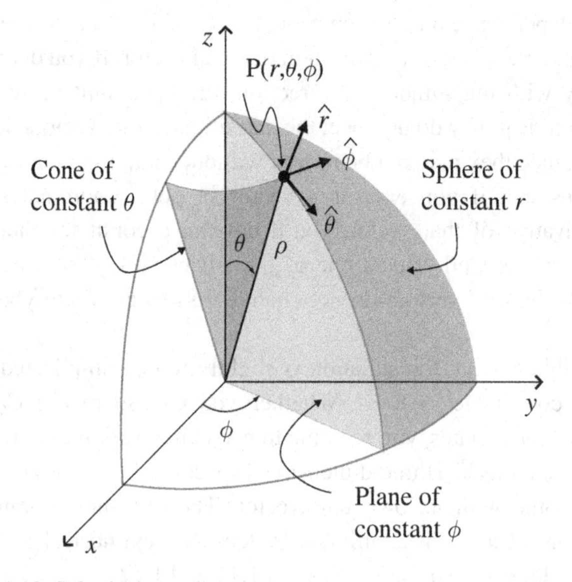

Figure 1.12 Spherical coordinates.

$$\begin{aligned} r &= \sqrt{x^2 + y^2} \\ \phi &= \arctan\left(\frac{y}{x}\right) \\ z &= z. \end{aligned} \tag{1.12}$$

And if you have the values of r, ϕ, and z, you can find x, y, and z using

$$\begin{aligned} x &= r\cos(\phi) \\ y &= r\sin(\phi) \\ z &= z. \end{aligned} \tag{1.13}$$

A vector at the point P is specified in cylindrical coordinates in terms of three mutually perpendicular components with unit vectors perpendicular to the cylinder of radius r, perpendicular to the plane through the z-axis at angle ϕ, and perpendicular to the xy plane at distance z. As in the Cartesian case, each cylindrical coordinate unit vector points in the direction in which that parameter is increasing, so $\hat{r}$ points in the direction of increasing r, $\hat{\phi}$ points in the direction of increasing ϕ, and $\hat{z}$ points in the direction of increasing z. The unit vectors $(\hat{r}, \hat{\phi}, \hat{z})$ form a right-handed set, so if you point the fingers of your right hand along $\hat{r}$ and push it into $\hat{\phi}$ with your right palm, your right thumb will show you the direction of $\hat{z}$.

The following equations relate the Cartesian to the cylindrical unit vectors:

$$\begin{aligned}
\hat{r} &= \cos(\phi)\hat{\imath} + \sin(\phi)\hat{\jmath} \\
\hat{\phi} &= -\sin(\phi)\hat{\imath} + \cos(\phi)\hat{\jmath} \\
\hat{z} &= \hat{z}.
\end{aligned} \tag{1.14}$$

In spherical coordinates a point P is specified by r, θ, ϕ where r represents the distance from the origin, θ is the angle measured from the z-axis toward the xy plane, and ϕ is the angle measured from the x-axis (or xz plane) to the constant-ϕ plane containing point P. With the z-axis up, θ is sometimes called the zenith angle and ϕ the azimuth angle. You can determine the spherical coordinates r, θ, and ϕ, from x, y, and z using the following equations:

$$\begin{aligned}
r &= \sqrt{x^2 + y^2 + z^2} \\
\theta &= \arccos\left(\frac{z}{\sqrt{x^2 + y^2 + z^2}}\right) \\
\phi &= \arctan\left(\frac{y}{x}\right).
\end{aligned} \tag{1.15}$$

And you can find x, y, and z from r, θ, and ϕ using:

$$\begin{aligned}
x &= r\sin(\theta)\cos(\phi) \\
y &= r\sin(\theta)\sin(\phi) \\
z &= r\cos(\theta).
\end{aligned} \tag{1.16}$$

In spherical coordinates, a vector at the point P is specified in terms of three mutually perpendicular components with unit vectors perpendicular to the sphere of radius r, perpendicular to the plane through the z-axis at angle ϕ, and perpendicular to the cone of angle θ. The unit vectors $(\hat{r}, \hat{\theta}, \hat{\phi})$ form a right-handed set, and are related to the Cartesian unit vectors as follows:

$$\begin{aligned}
\hat{r} &= \sin(\theta)\cos(\phi)\hat{\imath} + \sin(\theta)\sin(\phi)\hat{\jmath} + \cos(\theta)\hat{k} \\
\hat{\theta} &= \cos(\theta)\cos(\phi)\hat{\imath} + \cos(\theta)\sin(\phi)\hat{\jmath} - \sin(\theta)\hat{k} \\
\hat{\phi} &= -\sin(\phi)\hat{\imath} + \cos(\phi)\hat{\jmath}.
\end{aligned} \tag{1.17}$$

You may be asking yourself "Do I really need all these different unit vectors?" Well, *need* may be a bit strong, but your life will certainly be easier if you're trying to describe motion along a line of constant longitude on a spherical planet (the $\hat{\theta}$ direction) or the direction of a magnetic field around a current-carrying wire (the $\hat{\phi}$ direction). You'll find some examples of that in the problems at the end of this chapter.

1.6 Basis vectors

If you think about the unit vectors $\hat{\imath}$, $\hat{\jmath}$, and $\hat{k}$ and vector components such as $A_x\hat{\imath}$, $A_y\hat{\jmath}$, and $A_z\hat{k}$, you may realize that *any* vector in our three-dimensional Cartesian coordinate system can be made up of three components, each one telling you how many steps to take in the direction of one of the coordinate axes. Since those steps may be large or small, in the positive or negative direction, you can reach any point in the space containing these vectors. Little wonder, then, that $\hat{\imath}$, $\hat{\jmath}$, and $\hat{k}$ are one example of "basis vectors" in this space; combined with appropriate magnitudes, they form the *basis* of any vector in the space.

And you don't need to use only these particular vectors to make up any vector in this space – you can easily imagine using three vectors that are twice as long as the unit vectors $\hat{\imath}$, $\hat{\jmath}$, and $\hat{k}$, as shown in Figure 1.13(a). Although the vector components would change if you switched to these longer basis vectors, you'd have no trouble using them to make up any vector within the space. Specifically, if the unit vectors were twice as long, the values of A_x, A_y, and A_z would have to be only half as big to reach a given point in space.

You might even think of using three non-orthogonal, non-unit vectors such as the vectors $\vec{e}_1$, $\vec{e}_2$, and $\vec{e}_3$ in Figure 1.13(b) as your basis vectors. Of course, if you were to select three coplanar vectors (that is, vectors lying in the same plane), you'd quickly find that scaling and combining those vectors allows you

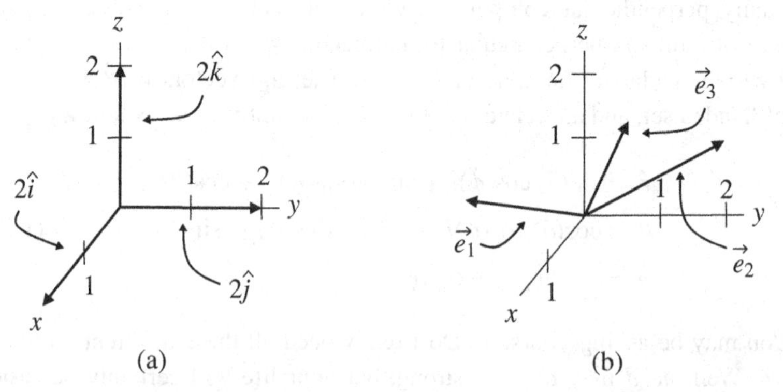

Figure 1.13 Alternative basis vectors.

to reach any point within that plane, but all points outside the plane would be unreachable. But as long as one of the three vectors is not coplanar with the other two, then appropriate scaling and combining will get you to any point in the space, and the vectors $\vec{e}_1$, $\vec{e}_2$, and $\vec{e}_3$ form a perfectly usable basis set (mathematicans say that they "span" the vector space).

You can ensure that three vectors are not coplanar by requiring them to be "linearly independent," which means that no two of the vectors may be scaled and combined to give the third, and no two are collinear (that is, lying along the same line or parallel to one another). This is often stated as the requirement that the only way to scale and combine the three vectors and get zero as the result is to scale each of the vectors by zero. In other words, for three linearly independent vectors $\vec{e}_1$, $\vec{e}_2$, and $\vec{e}_3$, the equation

$$A\vec{e}_1 + B\vec{e}_2 + C\vec{e}_3 = 0 \tag{1.18}$$

can only be true if A = B = C = 0.

So as long as you pick three linearly independent vectors, you have a viable set of basis vectors. And if you choose three non-coplanar vectors $\vec{e}_1$, $\vec{e}_2$, and $\vec{e}_3$ of non-unit length, it's quite simple to form unit vectors from these vectors. Since dividing a vector by a positive scalar changes its length but not its direction, you simply divide each vector by its magnitude:

$$\begin{aligned}\hat{e}_1 &= \frac{\vec{e}_1}{|\vec{e}_1|} \\ \hat{e}_2 &= \frac{\vec{e}_2}{|\vec{e}_2|} \\ \hat{e}_3 &= \frac{\vec{e}_3}{|\vec{e}_3|}.\end{aligned} \tag{1.19}$$

The concepts described in this section may be used to construct an infinite number of bases, but the most common are the "orthonormal" bases such as $\hat{\imath}$, $\hat{\jmath}$, and $\hat{k}$. These bases are called "ortho" because they're orthogonal (perpendicular to one another) and "normal" because they are normalized to a magnitude of one. Orthonormal bases will get you through the majority of problems you're likely to face.

One last fact about basis vectors in various coordinate systems will serve you very well if you study physics and engineering beyond the basic level, especially if your studies include the tensors discussed in Chapters 4 through 6. That fact is this: basis vectors that point along the axes of one coordinate system may be described in another coordinate system using partial derivatives.[8] Specifically, imagine that you're converting from spherical to rectangular coordinates. The basis vector along the original spherical (r) axis can be written in the Cartesian (x, y, and z) system as

$$\begin{aligned}\vec{e}_r &= \frac{\partial x}{\partial r}\hat{\imath} + \frac{\partial y}{\partial r}\hat{\jmath} + \frac{\partial z}{\partial r}\hat{k} \\ &= \sin\theta\cos\phi\;\hat{\imath} + \sin\theta\sin\phi\;\hat{\jmath} + \cos\theta\;\hat{k}.\end{aligned}$$

Likewise, the $\vec{e}_\theta$ and $\vec{e}_\phi$ basis vectors can be written as

$$\begin{aligned}\vec{e}_\theta &= \frac{\partial x}{\partial \theta}\hat{\imath} + \frac{\partial y}{\partial \theta}\hat{\jmath} + \frac{\partial z}{\partial \theta}\hat{k} \\ &= r\cos\theta\cos\phi\;\hat{\imath} + r\cos\theta\sin\phi\;\hat{\jmath} - r\sin\theta\;\hat{k}, \\ \vec{e}_\phi &= \frac{\partial x}{\partial \phi}\hat{\imath} + \frac{\partial y}{\partial \phi}\hat{\jmath} + \frac{\partial z}{\partial \phi}\hat{k} \\ &= -r\sin\theta\sin\phi\;\hat{\imath} + r\sin\theta\cos\phi\;\hat{\jmath}.\end{aligned}$$

Notice that these basis vectors are not all unit vectors (because their magnitudes are not all equal to one), nor do they all have the same dimensions ($\vec{e}_r$ is dimensionless, but $\vec{e}_\theta$ and $\vec{e}_\phi$ have dimensions of length). Neither of these characteristics disqualifies these as basis vectors, and you can always turn them into unit vectors by dividing by their magnitudes (take a look at the problems at the end of this chapter and their on-line solutions if you want to see how this works).

In general, if the coordinates of the original system are called x_1, x_2, and x_3 (these were r, θ, and ϕ in the example just discussed), and the coordinates of the new system are called x'_1, x'_2, and x'_3 (these were x, y, and z in the example), then the basis vectors along the original coordinate axes can be written in the new system as

$$\begin{aligned}\vec{e}_1 &= \frac{\partial x'_1}{\partial x_1}\vec{e}\,'_1 + \frac{\partial x'_2}{\partial x_1}\vec{e}\,'_2 + \frac{\partial x'_3}{\partial x_1}\vec{e}\,'_3, \\ \vec{e}_2 &= \frac{\partial x'_1}{\partial x_2}\vec{e}\,'_1 + \frac{\partial x'_2}{\partial x_2}\vec{e}\,'_2 + \frac{\partial x'_3}{\partial x_2}\vec{e}\,'_3, \\ \vec{e}_3 &= \frac{\partial x'_1}{\partial x_3}\vec{e}\,'_1 + \frac{\partial x'_2}{\partial x_3}\vec{e}\,'_2 + \frac{\partial x'_3}{\partial x_3}\vec{e}\,'_3.\end{aligned} \tag{1.20}$$

In other words, the partial derivatives $\frac{\partial x'_1}{\partial x_1}\vec{e}\,'_1$, $\frac{\partial x'_2}{\partial x_1}\vec{e}\,'_2$, and $\frac{\partial x'_3}{\partial x_1}\vec{e}\,'_3$ are the components of the first original (unprimed) basis vector expressed in the new (primed)

[8] If you're not familiar with partial derivatives or need a refresher, you'll find one in the next chapter.

coordinate system. For this reason, you'll find that some authors define basis vectors in terms of partial derivatives.

These relationships will prove to be extremely valuable in the study of coordinate-system transformation and tensor analysis, so file them away if your studies include those topics.

1.7 Chapter 1 problems

1.1 (a) If $|\vec{B}| = 18$ m and $\vec{B}$ points along the negative x-axis, what are B_x and B_y?
(b) If $C_x = -3$ m/s and $C_y = 5$ m/s, find the magnitude of $\vec{C}$ and the angle that $\vec{C}$ makes with the positive x-axis.

1.2 Vector $\vec{A}$ has magnitude of 11 m/s^2 and makes an angle of 65 degrees with the positive x-axis, and vector $\vec{B}$ has Cartesian components $B_x = 4$ m/s^2 and $B_y = -3$ m/s^2. If vector $\vec{C} = \vec{A} + \vec{B}$,
(a) Find the x- and y-components of $\vec{C}$;
(b) What are the magnitude and direction of $\vec{C}$?

1.3 Imagine that the y-axis points north and the x-axis points east.
(a) If you travel a distance $r = 22$ km in a straight line from the origin in a direction 35 degrees south of west, what is your position in Cartesian (x, y) coordinates?
(b) If you travel 6 miles due south from the origin and then turn west and travel 2 miles, how far from the origin and in what direction is your final position?

1.4 What are the x- and y-components of the polar unit vectors $\hat{r}$ and $\hat{\theta}$ when
(a) $\theta = 180$ degrees?
(b) $\theta = 45$ degrees?
(c) $\theta = 215$ degrees?

1.5 Cylindrical coordinates
(a) If $r = 2$ meters, $\phi = 35$ degrees, and $z = 1$ meter, what are x, y, and z?
(b) If $(x, y, z) = (3, 2, 4)$ meters, what are (r, ϕ, z)?

1.6 (a) In cylindrical coordinates, show that $\hat{r}$ points along the x-axis if $\phi = 0$.
(b) In what direction is $\hat{\phi}$ if $\phi = 90$ degrees?

1.7 (a) In spherical coordinates, find x, y, and z if $r = 25$ meters, $\theta = 35$ degrees, and $\phi = 110$ degrees.
(b) Find (r, θ, ϕ) if $(x, y, z) = (8, 10, 15)$ meters.

1.8 (a) For spherical coordinates, show that $\hat{\theta}$ points along the negative z-axis if $\theta = 90$ degrees.

(b) If ϕ also equals 90 degrees, in what direction are $\hat{r}$ and $\hat{\phi}$?

1.9 As you can read in Chapter 3, the magnetic field around a long, straight wire carrying a steady current I is given in spherical coordinates by the expression $\vec{B} = \frac{\mu_0 I}{2\pi R}\hat{\phi}$, where μ_0 is a constant and R is the perpendicular distance from the wire to the observation point. Find an expression for $\vec{B}$ in Cartesian coordinates.

1.10 If $\vec{e}_1 = 5\hat{\imath} - 3\hat{\jmath} + 2\hat{k}$, $\vec{e}_2 = \hat{\jmath} - 3\hat{k}$, and $\vec{e}_3 = 2\hat{\imath} + \hat{\jmath} - 4\hat{k}$, what are the unit vectors $\hat{e}_1$, $\hat{e}_2$, and $\hat{e}_3$?

2
Vector operations

If you were tracking the main ideas of Chapter 1, you should realize that vectors are representations of physical quantities – they're mathematical tools that help you visualize and describe a physical situation. In this chapter, you can read about a variety of ways to use those tools to solve problems. You've already seen how to add vectors and how to multiply vectors by a scalar (and why such operations are useful); this chapter contains many other "vector operations" through which you can combine and manipulate vectors. Some of these operations are simple and some are more complex, but each will prove useful in solving problems in physics and engineering. The first section of this chapter explains the simplest form of vector multiplication: the scalar product.

2.1 Scalar product

Why is it worth your time to understand the form of vector multiplication called the scalar or "dot" product? For one thing, forming the dot product between two vectors is very useful when you're trying to find the projection of one vector onto another. And why might you want to do that? Well, you may be interested in knowing how much work is done by a force acting on an object. The first instinct of many students is to think of work as "force times distance" (which is a reasonable starting point). But if you've ever taken a course that went a bit deeper than the introductory level, you may remember that the definition of work as force times distance applies only to the special case in which the force points in exactly the same direction as the displacement of the object. In the more general case in which the force acts at some angle to the direction of the displacement, you have to find the component of the force along the displacement. That's one example of exactly what the dot product can do for you, and you'll find more in the problems at the end of this chapter.

How do you go about computing the dot product between two vectors? Well, if you know the Cartesian components of each vector (call the vectors $\vec{A}$ and $\vec{B}$), you can use

$$\vec{A} \circ \vec{B} = A_x B_x + A_y B_y + A_z B_z. \tag{2.1}$$

Or if you know the angle θ between the vectors,

$$\vec{A} \circ \vec{B} = |\vec{A}||\vec{B}| \cos\theta, \tag{2.2}$$

where $|\vec{A}|$ and $|\vec{B}|$ represent the magnitude (length) of the vectors $\vec{A}$ and $\vec{B}$.[1] Note that the dot product between two vectors gives a scalar result (just a single value, no direction).

To grasp the physical significance of the dot product, consider vectors $\vec{A}$ and $\vec{B}$ which differ in direction by angle θ, as shown in Figure 2.1a. For these vectors, the projection of $\vec{A}$ onto the direction of $\vec{B}$ is $|\vec{A}| \cos(\theta)$, as shown in Figure 2.1b. Multiplying this projection by the length of $\vec{B}$ gives $|\vec{A}||\vec{B}| \cos(\theta)$. Thus the dot product $\vec{A} \circ \vec{B}$ represents the projection of $\vec{A}$ onto the direction of $\vec{B}$ multiplied by the length of $\vec{B}$. The scalar result of this operation is exactly the same as the result of finding the projection of $\vec{B}$ onto the direction of $\vec{A}$ and then multiplying that value by the length of $\vec{A}$. Hence the order of the two vectors in the dot product is irrelevant; $\vec{A} \circ \vec{B}$ gives the same result as $\vec{B} \circ \vec{A}$.

The scalar product can be particularly useful when one of the vectors in the product is a unit vector. That's because the length of a unit vector is by definition equal to one, so a scalar product such as $\vec{A} \circ \hat{k}$ finds the projection of vector $\vec{A}$ onto the direction of $\hat{k}$ (the z-direction) multiplied by the magnitude of $\hat{k}$ (which is one). Thus to find the component of any vector in a given direction, you can simply form the dot product between that vector and the unit vector in

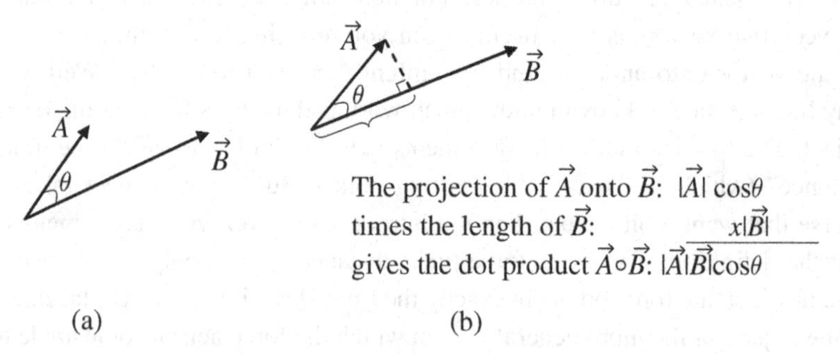

Figure 2.1 Two vectors and their scalar product.

[1] The equivalence between Equations 2.1 and 2.2 is demonstrated in the problems at the end of this chapter.

the desired direction. It's quite likely you'll come across problems in physics and engineering in which you have a vector ($\vec{A}$) and you wish to know the component of that vector that's perpendicular to a specified surface; if you know the unit normal vector ($\hat{n}$) for the surface, the scalar product $\vec{A} \circ \hat{n}$ gives you that perpendicular component of $\vec{A}$.

The scalar product is also useful in finding the angle between two vectors. To understand how that works, consider the two expressions for the dot product given in Eqs. 2.1 and 2.2. Since

$$\vec{A} \circ \vec{B} = |\vec{A}||\vec{B}| \cos\theta = A_x B_x + A_y B_y + A_z B_z, \tag{2.3}$$

then dividing both sides by the product of the magnitudes of $\vec{A}$ and $\vec{B}$ gives

$$\cos(\theta) = \frac{A_x B_x + A_y B_y + A_z B_z}{|\vec{A}||\vec{B}|}$$

or

$$\theta = \arccos\left(\frac{A_x B_x + A_y B_y + A_z B_z}{|\vec{A}||\vec{B}|}\right). \tag{2.4}$$

So if you wish to find the angle between two vectors $\vec{A} = 5\hat{\imath} - 2\hat{\jmath} + 4\hat{k}$ and $\vec{B} = 3\hat{\imath} + \hat{\jmath} + 7\hat{k}$, you can use Eq. 2.4 to find

$$\begin{aligned}\theta &= \arccos\left(\frac{(5)(3) + (-2)(1) + (4)(7)}{\sqrt{(5)^2 + (-2)^2 + (4)^2}\sqrt{(3)^2 + (1)^2 + (7)^2}}\right)\\ &= \arccos\left(\frac{41}{\sqrt{45}\sqrt{59}}\right)\\ &= 37.3^\circ.\end{aligned}$$

One final note about the scalar product: any unit vector dotted with itself gives a result of 1 (since, for example, $\hat{\imath} \circ \hat{\imath} = |\hat{\imath}||\hat{\imath}| \cos(0^\circ) = (1)(1)(1) = 1$), and the dot product between two different orthogonal unit vectors gives a result of zero (since, for example, $\hat{\imath} \circ \hat{\jmath} = |\hat{\imath}||\hat{\jmath}| \cos(90^\circ) = (1)(1)(0) = 0$).

2.2 Cross product

Another way to multiply two vectors is to form the "cross product" between them. Unlike the dot product, which gives a scalar result, the cross product results in another vector. Why bother learning this form of vector multiplication? One reason is that the cross product is just what you need when you're trying to find the result of certain physical processes, such as applying a force at the end of a lever arm or firing a charged particle into a magnetic field.

Computing the cross product between two vectors is only slightly more complicated than finding the dot product. If you know the Cartesian components of both vectors, the cross product is given by

$$\begin{aligned}\vec{A}\times\vec{B} &= (A_yB_z - A_zB_y)\hat{\imath}\\ &\quad + (A_zB_x - A_xB_z)\hat{\jmath}\\ &\quad + (A_xB_y - A_yB_x)\hat{k},\end{aligned} \tag{2.5}$$

which can be written as

$$\vec{A}\times\vec{B} = \begin{vmatrix} \hat{\imath} & \hat{\jmath} & \hat{k} \\ A_x & A_y & A_z \\ B_x & B_y & B_z \end{vmatrix}. \tag{2.6}$$

If you haven't seen determinants before and you need some help getting from Eq. 2.6 to Eq. 2.5, you can find an explanation of how this works on the book's website.

The direction of the vector formed by the cross product of $\vec{A}$ and $\vec{B}$ is perpendicular to both $\vec{A}$ and $\vec{B}$ (that is, perpendicular to the plane containing both $\vec{A}$ and $\vec{B}$), as shown in Figure 2.2. Of course, there are two directions perpendicular to this plane, so how do you know which one corresponds to the direction of $\vec{A}\times\vec{B}$? The answer is provided by the "right-hand rule," which you can invoke by opening your right hand and making your thumb perpendicular to the direction of your fingers in the plane of your palm. Now imagine using your right palm and fingers to push the first vector ($\vec{A}$ in this case) into the direction of the second vector ($\vec{B}$ in this case) through the smallest angle. As you push, your thumb shows you the direction of the cross product.[2]

A very important difference between the dot product and the cross product is that the order of the vectors is irrelevant for the dot product but matters greatly for the cross product. You can see this by imagining the cross product $\vec{B}\times\vec{A}$ in Figure 2.2. In order to push vector $\vec{B}$ into vector $\vec{A}$ with your right palm, you'd have to turn your hand upside-down (that is, with your thumb pointing down). And since your thumb shows you the direction of the cross product, you can see that $\vec{B}\times\vec{A}$ points in the opposite direction from $\vec{A}\times\vec{B}$. That means that

$$\vec{A}\times\vec{B} = -\vec{B}\times\vec{A}, \tag{2.7}$$

[2] Some people find it easier to imagine aligning the fingers of your (open) right hand with the direction of the first vector, and then curling your fingers toward the second vector. Or you can point your right index finger in the direction of the first vector and your right middle finger in the direction of the second vector. Whether you use the pushing, curling, or pointing approach, your right thumb shows you the direction of the cross product.

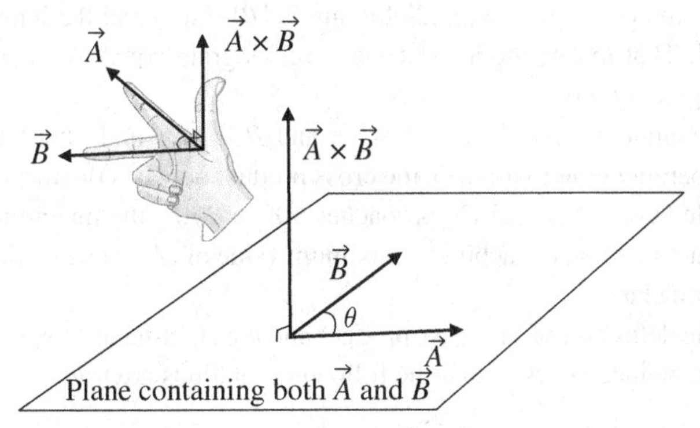

Figure 2.2 Direction of the cross product $\vec{A} \times \vec{B}$.

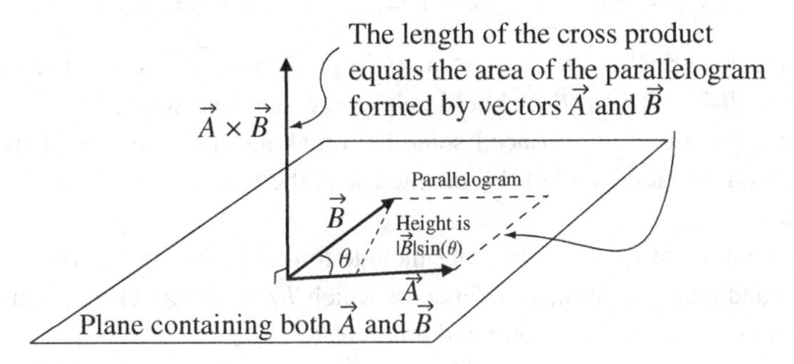

Figure 2.3 The cross product as area.

since the negative of a vector is just a vector of the same magnitude in the opposite direction. A quick method of computing the magnitude of the cross product is to use

$$|\vec{A} \times \vec{B}| = |\vec{A}||\vec{B}|\sin(\theta), \tag{2.8}$$

where $|\vec{A}|$ is the magnitude of $\vec{A}$, $|\vec{B}|$ is the magnitude of $\vec{B}$, and θ is the angle between $\vec{A}$ and $\vec{B}$.[3]

One way to picture the length and direction of the cross product is illustrated in Figure 2.3. Just as the dot product involves the projection of one vector onto another, the cross product also has a geometrical interpretation. In this case, the magnitude of the cross product between two vectors is proportional to the area of the parallelogram formed with those two vectors as adjacent sides. As you may recall, the area of a parallelogram is just its base times its height, and

[3] The equivalence of Eq. 2.8 and the magnitude of the expression in Eq. 2.5 is demonstrated in the problems at the end of this chapter.

in this case the height of the parallelogram is $|\vec{B}|\sin(\theta)$ and the length of the base is $|\vec{A}|$. That makes the area of the parallelogram equal to $|\vec{A}||\vec{B}|\sin(\theta)$, exactly as given in Eq. 2.8.

So if the angle between two vectors $\vec{A}$ and $\vec{B}$ is zero or 180° (that is, if $\vec{A}$ and $\vec{B}$ are parallel or antiparallel), the cross product between them is zero. And as the angle between $\vec{A}$ and $\vec{B}$ approaches 90° or 270°, the magnitude of the cross product increases, reaching a maximum value of $|\vec{A}||\vec{B}|$ when the vectors are perpendicular.

Using the definition of the cross product and the right-hand rule, you should be able to convince yourself that the following relations are true:

$$\begin{array}{lll} \hat{\imath}\times\hat{\imath}=0 & \hat{\imath}\times\hat{\jmath}=\hat{k} & \hat{\jmath}\times\hat{\imath}=-\hat{k} \\ \hat{\jmath}\times\hat{\jmath}=0 & \hat{\jmath}\times\hat{k}=\hat{\imath} & \hat{k}\times\hat{\jmath}=-\hat{\imath} \\ \hat{k}\times\hat{k}=0 & \hat{k}\times\hat{\imath}=\hat{\jmath} & \hat{\imath}\times\hat{k}=-\hat{\jmath}. \end{array} \tag{2.9}$$

Applying these relations term-by-term to the product of $\vec{A} = A_x\hat{\imath}+A_y\hat{\jmath}+A_z\hat{k}$ and $\vec{B} = B_x\hat{\imath} + B_y\hat{\jmath} + B_z\hat{k}$ should help you understand where Eqs. 2.6 and 2.5 come from (and if you need some help making that work out, there's a problem on this at the end of this chapter, with the full solution on the book's website).

Applications of the cross product include torque problems (in which $\vec{\tau} = \vec{r} \times \vec{F}$) and magnetic force problems (in which $\vec{F}_B = q\vec{v} \times \vec{B}$); you can find examples of these in the chapter-end problems.

2.3 Triple scalar product

Once you understand the dot product and cross product described in the previous two sections, you may be wondering if it's possible to combine these two vector operations. Happily, it's not only possible, it's actually *useful* to do so. After all, you can define all the mathematical operations you'd like, but unless those operations result in something that you can apply to solve problems, you'd have to leave them in the "curiosity" file. You've seen how the dot product finds employment when projections of vectors onto specified directions are needed and when work is to be calculated, and how the cross product can be called into action when torques and magnetic forces are at play. But does it make sense to combine the dot and cross product operations in a manner such as $\vec{A}\circ(\vec{B}\times\vec{C})$? Yes it does.[4] This is called the "triple scalar product" or "scalar triple product" and it has several useful applications.

[4] But $(\vec{A}\circ\vec{B})\times\vec{C}$ makes no sense, since $(\vec{A}\circ\vec{B})$ gives a scalar, and you can't cross that scalar into $\vec{C}$.

The mathematics of this operation are straightforward; you know that

$$\begin{aligned}\vec{B} \times \vec{C} = (B_y C_z - B_z C_y)\hat{\imath} \\ + (B_z C_x - B_x C_z)\hat{\jmath} \\ + (B_x C_y - B_y C_x)\hat{k},\end{aligned} \tag{2.10}$$

and from Eq. 2.1 you also know that

$$\vec{A} \circ \vec{B} = A_x B_x + A_y B_y + A_z B_z,$$

so combining the dot and cross product gives

$$\begin{aligned}\vec{A} \circ \left(\vec{B} \times \vec{C}\right) = A_x(B_y C_z - B_z C_y) \\ + A_y(B_z C_x - B_x C_z) \\ + A_z(B_x C_y - B_y C_x).\end{aligned} \tag{2.11}$$

A handy way to write this is

$$\vec{A} \circ \left(\vec{B} \times \vec{C}\right) = \begin{vmatrix} A_x & A_y & A_z \\ B_x & B_y & B_z \\ C_x & C_y & C_z \end{vmatrix}. \tag{2.12}$$

One geometrical interpretation of the triple scalar product can be understood with the help of Figure 2.4. In this figure, vectors $\vec{A}$, $\vec{B}$, and $\vec{C}$ represent the sides of a parallelepiped. The area of the base of this parallelepiped is $|\vec{B} \times \vec{C}|$, as in Figure 2.3, and its height is equal to $|\vec{A}|\cos(\phi)$, where ϕ is the angle between $\vec{A}$ and the direction of $\vec{B} \times \vec{C}$. That means that the volume of the parallelepiped (the height times the area of the base) must be $|\vec{A}|\cos(\phi)(|\vec{B} \times \vec{C}|)$. Writing this as $|\vec{A}||\vec{B} \times \vec{C}|\cos(\phi)$ should help you see that this has the same form as the definition of the dot product in Eq. 2.2 and is therefore just $\vec{A} \circ (\vec{B} \times \vec{C})$.

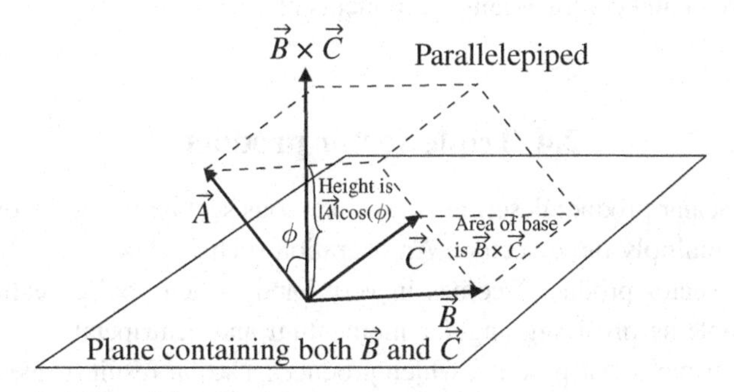

Figure 2.4 The triple scalar product as volume.

Hence the triple scalar product $\vec{A} \circ (\vec{B} \times \vec{C})$ may be interpreted as the volume of the parallelepiped formed by vectors $\vec{A}$, $\vec{B}$, and $\vec{C}$. You should note that the triple product will give a positive result so long as the vectors $\vec{A}$, $\vec{B}$, and $\vec{C}$ form a right-handed system (that is, pushing $\vec{A}$ into $\vec{B}$ with the palm of your right hand gives a direction onto which $\vec{C}$ projects in a positive sense (likewise for pushing $\vec{B}$ into $\vec{C}$ and pushing $\vec{C}$ into $\vec{A}$).

Seeing the relationship between the triple scalar product of three vectors and the volume formed by those vectors makes it easy to understand why the triple scalar product may be used as a test to determine whether three vectors are coplanar (that is, whether all three lie in the same plane). Just imagine how the parallelepiped in Figure 2.4 would look if vectors $\vec{A}$, $\vec{B}$, and $\vec{C}$ were all in the same plane. In that case, the height of the parallelepiped would be zero and the projection of $\vec{A}$ onto the direction of $\vec{B} \times \vec{C}$ would be zero, which means the triple product $\vec{A} \circ (\vec{B} \times \vec{C})$ would have to be zero. Stated another way, if the projection of $\vec{A}$ onto the direction of $\vec{B} \times \vec{C}$ is not zero, then $\vec{A}$ cannot lie in the same plane as $\vec{B}$ and $\vec{C}$. Thus

$$\vec{A} \circ (\vec{B} \times \vec{C}) = 0 \tag{2.13}$$

is both a necessary and a sufficient condition for vectors $\vec{A}$, $\vec{B}$, and $\vec{C}$ to be coplanar.

Equating $\vec{A} \circ (\vec{B} \times \vec{C})$ to the volume of the parallelepiped formed by vectors $\vec{A}$, $\vec{B}$, and $\vec{C}$ should also help you see that any cyclic permutation of the vectors (such as $\vec{B} \circ (\vec{C} \times \vec{A})$ or $\vec{C} \circ (\vec{A} \times \vec{B})$) gives the same result for the triple scalar product, since the volume of the parallelepiped is the same in each of these cases. Some authors describe this as the ability to interchange the dot and the cross without affecting the result (since $(\vec{A} \times \vec{B}) \circ \vec{C}$ is the same as $\vec{C} \circ (\vec{A} \times \vec{B})$).

One application in which the triple scalar product finds use is the determination of reciprocal vectors, as explained in the sections in Chapter 4 dealing with covariant and contravariant components of vectors.

2.4 Triple vector product

The triple scalar product described in the previous section is not the only useful way to multiply three vectors. An operation such as $\vec{A} \times (\vec{B} \times \vec{C})$ (called the "triple vector product") comes in very handy when you're dealing with certain problems involving angular momentum and centripetal acceleration. Unlike the triple scalar product, which produces a scalar result (since the second operation is a dot product), the triple vector product yields a vector result (since both operations are cross products). You should note that $\vec{A} \times (\vec{B} \times \vec{C})$ is not the same as $(\vec{A} \times \vec{B}) \times \vec{C}$; the location of the parentheses matters greatly in the triple vector product. The triple vector product is somewhat tedious to calculate by brute force, but thankfully a simplified expression exists:

$$\vec{A} \times (\vec{B} \times \vec{C}) = \vec{B}(\vec{A} \circ \vec{C}) - \vec{C}(\vec{A} \circ \vec{B}). \tag{2.14}$$

After all the previous discussion of the various ways in which vectors can be multiplied, you can be forgiven for thinking that the right side of this equation looks a bit strange, with no circle or cross between $\vec{B}$ and $\vec{A} \circ \vec{C}$ or between $\vec{C}$ and $\vec{A} \circ \vec{B}$. Just remember that $\vec{A} \circ \vec{C}$ and $\vec{A} \circ \vec{B}$ are scalars, so the expressions in parentheses in Eq. 2.14 are simply scalar multipliers of vectors $\vec{B}$ and $\vec{C}$. Does this mean that the result of the operation $\vec{A} \times (\vec{B} \times \vec{C})$ is a vector that is some linear combination of the second and third vectors in the triple product? That's exactly what it means, as you can see by considering Figure 2.5.

In this figure, you can see the vector $\vec{B} \times \vec{C}$ pointing straight up, perpendicular to the plane containing vectors $\vec{B}$ and $\vec{C}$. Now imagine forming the cross product of vector $\vec{A}$ with vector $\vec{B} \times \vec{C}$ by pushing $\vec{A}$ into the direction of $\vec{B} \times \vec{C}$ with the palm of your right hand. The result of this operation, labelled vector $\vec{A} \times (\vec{B} \times \vec{C})$, is back in the plane containing vectors $\vec{B}$ and $\vec{C}$. To understand why this is true, consider the fact that the vector that results from the operation $\vec{B} \times \vec{C}$ must be perpendicular to the plane containing $\vec{B}$ and $\vec{C}$. If you now cross $\vec{A}$ into that vector, the resulting vector must be perpendicular to both $\vec{A}$ and to $(\vec{B} \times \vec{C})$, which puts it back in the plane containing vectors $\vec{B}$ and $\vec{C}$. And if the vector result of the operation $\vec{A} \times (\vec{B} \times \vec{C})$ is in the same plane as vectors $\vec{B}$ and $\vec{C}$, then it must be a linear combination of those two vectors.

You can remember Eq. 2.14 as the "BAC minus CAB" rule so long as you remember to write the members of the triple product in the correct sequence

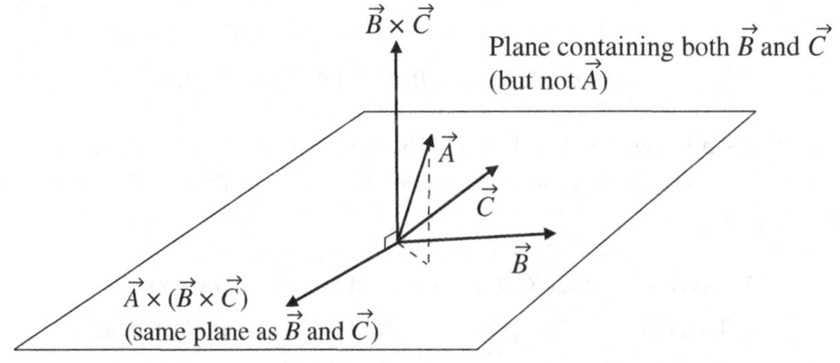

Figure 2.5 Vectors involved in the triple vector product $\vec{A} \times (\vec{B} \times \vec{C})$.

$(\vec{A}, \vec{B}, \vec{C})$ with the parentheses around the last two vectors. To see where this comes from, you can simply use the definition of the cross product (Eq. 2.6) to write

$$\vec{A} \times (\vec{B} \times \vec{C}) = \begin{vmatrix} \hat{\imath} & \hat{\jmath} & \hat{k} \\ A_x & A_y & A_z \\ (\vec{B} \times \vec{C})_x & (\vec{B} \times \vec{C})_y & (\vec{B} \times \vec{C})_z \end{vmatrix}. \tag{2.15}$$

And from Equation 2.5 you know that

$$\begin{aligned} \vec{B} \times \vec{C} = (B_y C_z - B_z C_y)\hat{\imath} \\ + (B_z C_x - B_x C_z)\hat{\jmath} \\ + (B_x C_y - B_y C_x)\hat{k}. \end{aligned} \tag{2.16}$$

Substituting these terms into Eq. 2.15 gives

$$\vec{A} \times (\vec{B} \times \vec{C}) = \begin{vmatrix} \hat{\imath} & \hat{\jmath} & \hat{k} \\ A_x & A_y & A_z \\ (B_y C_z - B_z C_y) & (B_z C_x - B_x C_z) & (B_x C_y - B_y C_x) \end{vmatrix}. \tag{2.17}$$

Multiplying this out looks ugly at first:

$$\begin{aligned} \vec{A} \times (\vec{B} \times \vec{C}) = {} & [A_y(B_x C_y - B_y C_x) - A_z(B_z C_x - B_x C_z)]\hat{\imath} \\ & + [A_z(B_y C_z - B_z C_y) - A_x(B_x C_y - B_y C_x)]\hat{\jmath} \\ & + [A_x(B_z C_x - B_x C_z) - A_y(B_y C_z - B_z C_y)]\hat{k}. \end{aligned} \tag{2.18}$$

But a little rearranging gives

$$\begin{aligned} \vec{A} \times (\vec{B} \times \vec{C}) = {} & (A_y C_y + A_z C_z)(B_x \hat{\imath}) - (A_y B_y + A_z B_z)(C_x \hat{\imath}) \\ & + (A_z C_z + A_x C_x)(B_y \hat{\jmath}) - (A_z B_z + A_x B_x)(C_y \hat{\jmath}) \\ & + (A_x C_x + A_y C_y)(B_z \hat{k}) - (A_x B_x + A_y B_y)(C_z \hat{k}), \end{aligned} \tag{2.19}$$

which still isn't pretty, but it does hold some promise. That promise can be realized by adding nothing to each row of Eq. 2.19. Nothing, that is, in the following form:

$A_x B_x C_x(\hat{\imath}) - A_x B_x C_x(\hat{\imath})$ Add this to the top row;

$A_y B_y C_y(\hat{\jmath}) - A_y B_y C_y(\hat{\jmath})$ Add this to the middle row;

$A_z B_z C_z(\hat{k}) - A_z B_z C_z(\hat{k})$ Add this to the bottom row.

These additions make Eq. 2.19 a good deal more friendly:

$$\begin{aligned}
&\vec{A} \times (\vec{B} \times \vec{C}) \\
&\quad = (A_x C_x + A_y C_y + A_z C_z)(B_x \hat{\imath}) - (A_x B_x + A_y B_y + A_z B_z)(C_x \hat{\imath}) \\
&\qquad + (A_x C_y + A_y C_y + A_z C_z)(B_y \hat{\jmath}) - (A_x B_x + A_y B_y + A_z B_z)(C_y \hat{\jmath}) \\
&\qquad + (A_x C_x + A_y C_y + A_z C_z)(B_z \hat{k}) - (A_x B_x + A_y B_y + A_z B_z)(C_z \hat{k}).
\end{aligned}$$

Or

$$\begin{aligned}
\vec{A} \times (\vec{B} \times \vec{C}) = {} & (A_x C_x + A_y C_y + A_z C_z)(B_x \hat{\imath} + B_y \hat{\jmath} + B_z \hat{k}) \\
& - (A_x B_x + A_y B_y + A_z B_z)(C_x \hat{\imath} + C_y \hat{\jmath} + C_z \hat{k}).
\end{aligned}$$

But $B_x \hat{\imath} + B_y \hat{\jmath} + B_z \hat{k}$ is just the vector $\vec{B}$, $C_x \hat{\imath} + C_y \hat{\jmath} + C_z \hat{k}$ is the vector $\vec{C}$, and the other two terms fit the definition of dot products (Eq. 2.1). Thus

$$\begin{aligned}
\vec{A} \times (\vec{B} \times \vec{C}) &= (\vec{A} \circ \vec{C})\vec{B} - (\vec{A} \circ \vec{B})\vec{C} \\
&= \vec{B}(\vec{A} \circ \vec{C}) - \vec{C}(\vec{A} \circ \vec{B}).
\end{aligned}$$

2.5 Partial derivatives

Once you understand the basic vector operations of dot, cross, and triple products, it's a small step to more advanced vector operations such as gradient, divergence, curl, and the Laplacian. But these are *differential* vector operations, so before you can make that step, it's important for you to understand the difference between ordinary derivatives and partial derivatives. This is worth your time and effort because differential vector operations have many applications in diverse areas of physics and engineering.

You probably first encountered ordinary derivatives when you learned how to find the slope of a line ($m = \frac{dy}{dx}$) or how to determine the speed of an object given its position as a function of time ($v_x = \frac{dx}{dt}$). Happily, partial derivatives are based on the same general concepts as ordinary derivatives, but extend those concepts to functions of multiple variables. And you should never have any doubt as to which kind of derivative you're dealing with, because ordinary derivatives are written as $\frac{d}{dx}$ or $\frac{d}{dt}$ and partial derivatives are written as $\frac{\partial}{\partial x}$ or $\frac{\partial}{\partial t}$.

As you may recall, ordinary derivatives come about when you're interested in the *change* of one variable with respect to another. For example, you may encounter a variable y which is a function of another variable x (which means that the value of y depends on the value of x). This can be written as $y = f(x)$, where y is called the "dependent variable" and x is called the "independent variable." The ordinary derivative of y with respect to x (written as $\frac{dy}{dx}$) tells you how much the value of y changes for a small change in the variable x. If you make a graph with y on the vertical axis and x on the horizontal axis, as in Figure 2.6, then the slope of the line between any two points (x_1, y_1) and (x_2, y_2) on the graph is simply $\frac{y_2-y_1}{x_2-x_1} = \frac{\Delta y}{\Delta x}$. That's because the slope is defined as "the rise over the run," and since the rise is Δy for a run Δx, the slope of the line between any two points must be $\frac{\Delta y}{\Delta x}$.

But if you look closely at the expanded region of Figure 2.6, you'll notice that the graph of y versus x has a slight curve between points (x_1, y_1) and (x_2, y_2), so the slope is actually *changing* in that interval. Thus the ratio $\frac{\Delta y}{\Delta x}$ can't represent the slope everywhere between those points. Instead, it represents the average slope over this interval, as suggested by the dashed line between points (x_1, y_1) and (x_2, y_2) (which by the mean value theorem does equal the slope somewhere in between these two points, but not necessarily in the middle). To represent the slope at a given point on the curve more precisely, all you have to do is to allow the "run" Δx to become very small. As Δx approaches zero, the difference between the dashed line and the curved line in Figure 2.6 becomes negligible. If you write the incremental run as dx and the (also incremental) rise as dy, then the slope at any point on the line can be written as $\frac{dy}{dx}$. This is the reasoning that equates the derivative of a function to the slope of the graph of that function.

Now imagine that you have a variable z that depends on two other variables, say x and y, so $z = f(x, y)$. One way to picture such a case is to visualize a surface in three-dimensional space, as in Figure 2.7. The height of this surface above the xy plane is z, which gets higher and lower at different values of x and y. And since the height z may change at a different rate in different directions, a single derivative will not generally be sufficient to characterize the total change in height as you move from one point to another. You can see

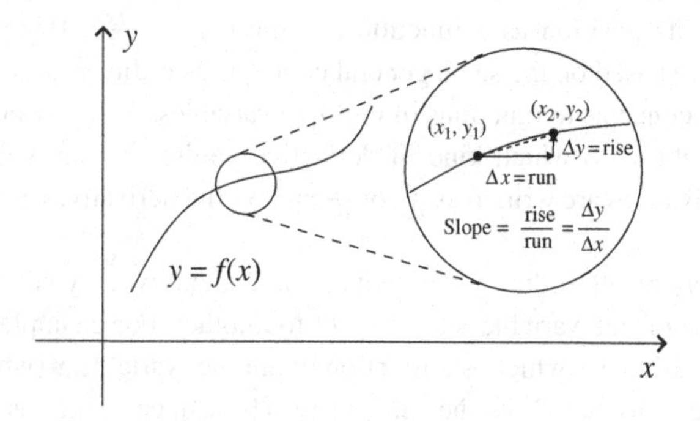

Figure 2.6 Slope of the line $y = f(x)$.

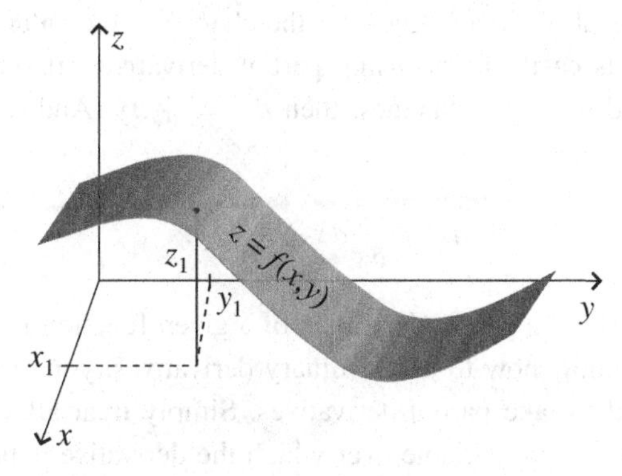

Figure 2.7 Surface in 3-D space ($z = f(x, y)$).

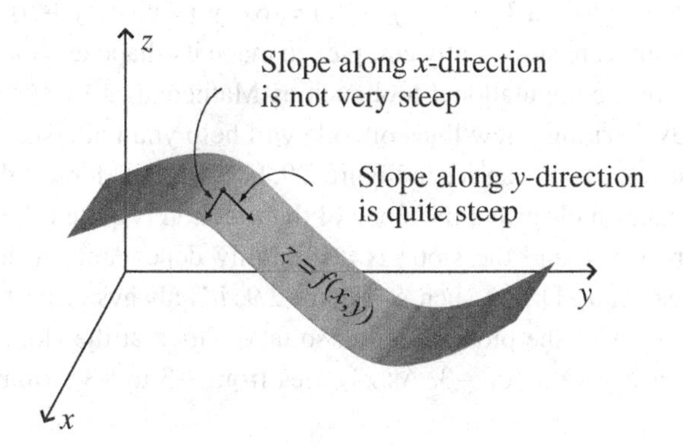

Figure 2.8 Surface in 3-D space ($z = f(x, y)$).

the height z changing at different rates in Figure 2.8; at the location shown in the figure, the slope of the surface is quite steep if you move in the direction of increasing y (while remaining at the same value of x), but the slope is almost zero if you move in the direction of increasing x (while holding your y-value constant).

This illustrates the usefulness of *partial* derivatives, which are derivatives formed by allowing one independent variable (such as x or y in Figure 2.8) to change while holding other independent variables constant. So the partial derivative $\frac{\partial z}{\partial x}$ represents the slope of the surface at a given location if you move *only along the x-direction* from that location, and the partial derivative $\frac{\partial z}{\partial y}$ represents the slope if you move *only along the y-direction*. You may find these partial derivatives written as $\frac{\partial z}{\partial x}|_y$ and $\frac{\partial z}{\partial y}|_x$, where the variables that appear in the subscript after the vertical line are held constant.

As you've probably already guessed, the change in the value of z as either x or y changes is easily found using partial derivatives. If only x changes, $dz = \frac{\partial z}{\partial x}dx$, and if only y changes, then $dz = \frac{\partial z}{\partial y}dy$. And if both x and y change, then

$$dz = \frac{\partial z}{\partial x}dx + \frac{\partial z}{\partial y}dy. \tag{2.20}$$

The process of taking a partial derivative of a given function is quite straightforward; if you know how to take ordinary derivatives, you already have the tools you'll need to take partial derivatives. Simply treat all variables (with the exception of the one variable over which the derivative is being taken) as constants, and take the derivative as you normally would. This is best explained using an example.

Consider a function such as $z = f(x, y) = 6x^2y+3x+5xy+10$. The terms of this polynomial are sufficiently complex to make its shape less than obvious, which is where a computational tool such as Mathematica or MATLAB can be very handy. Writing a few lines of code will help you understand how this function behaves, as you can see in Figure 2.9. Even a quick look at this warped little plane makes it clear that the slope of the function is quite different in the x- and y-directions, and the slope is also highly dependent on the location on the surface. In a 3D plot such as Figure 2.9, it's always easiest to see the slope at the edges of the plotted region, so take a look at the slope along the x-direction for a y value of -3. As x varies from -3 to +3 (while y is held

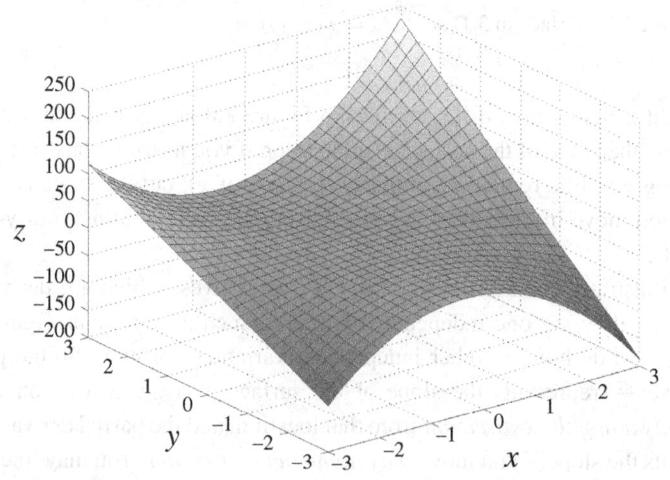

Figure 2.9 Plot of the function $z = f(x, y) = 6x^2y + 3x + 5xy + 10$ for $-3 \leq x \leq 3$ and $-3 \leq y \leq 3$.

constant at -3), the slope starts off positive and gets less steep as you move in the $+x$-direction from $x = -3$ toward $x = 0$. The slope then becomes zero somewhere near $x = 0$, then turns negative and becomes increasingly steep as x approaches $+3$. Doing the same quick analysis along the y-direction while holding x constant at -3 indicates that the slope is approximately constant and positive as y varies from -3 to $+3$.

Now that you have some idea of what to expect, you can take the partial derivative of $z = 6x^2y + 3x + 5xy + 10$ with respect to x simply by treating the variable y as a constant:

$$\frac{\partial z}{\partial x} = 12xy + 3 + 5y. \tag{2.21}$$

Likewise, the partial derivative with respect to y is found by holding x constant:

$$\frac{\partial z}{\partial y} = 6x^2 + 5x. \tag{2.22}$$

Before interpreting these derivative results, you may want to take a moment to make sure you understand why the process of taking the derivative of a function involves bringing down the exponent of the relevant variable and then subtracting one from that exponent (so $\frac{d(x^2)}{dx} = 2x$, for example). The answer is quite straightforward. Since the derivative represents the change in the function z as the independent variable x changes over a very small run, the formal definition for this derivative can be written as

$$\frac{dz}{dx} \equiv \lim_{\Delta x \to 0} \frac{z(x + \Delta x) - z(x)}{\Delta x}. \tag{2.23}$$

So in the case of $z = x^2$, you have

$$\frac{d(x^2)}{dx} \equiv \lim_{\Delta x \to 0} \frac{(x + \Delta x)^2 - x^2}{\Delta x}. \tag{2.24}$$

If you think about the term in the numerator, you'll see that this is $x^2 + 2x\Delta x + (\Delta x)^2 - x^2$, which is just $2x\Delta x + (\Delta x)^2$, and dividing this by Δx gives $2x + \Delta x$. But as Δx approaches zero, the Δx term becomes negligible, and this approaches $2x$. So where did the 2 come from? It's just the number of cross terms (that is, terms with the product of x and Δx) that result from raising $(x + \Delta x)$ to the second power. Had you been taking the derivative of x^3 with respect to x, you would have had three such cross terms. So you bring down the exponent because that's the number of cross terms that result from taking $x + \Delta x$ to that power. And why do you then subtract one from the exponent? Simply because when you take the *change* in the function z (that is, $(x + \Delta x)^2 - x^2$), the highest-power terms (x^2 in this case) cancel, leaving only terms of one lower power (x^1 in this case). It's a bit laborious, but the same analysis can be applied to show that $\frac{d(x^3)}{dx} = 3x^2$ and that $\frac{d(x^n)}{dx} = nx^{n-1}$.

So that's why you bring down the exponent and subtract one, but what does it mean when you take derivatives and get answers such as Eqs. 2.21 and 2.22? It simply means that the slope varies with direction and location on the surface z. So, for example, the slope along the x-direction at location (−3,2) is $12xy + 3 + 5y = 12(-3)(2) + 3 + 5(2) = -59$, while at the same location the slope along the y-direction is $6x^2 + 5x = 6[(-3)^2] + 5(-3) = 39$.

You can do a rough check on your calculated partial derivative in Eq. 2.21 by inserting the value of −3 for y to see that the slope of z at this value of y is $12(x)(-3) + 3 + 5(-3) = -36x - 12$. Thus as you move in the x-direction at $y = -3$, the slope should vary from +96 at $x = -3$, to zero at $x = -1/3$, and down to −120 at $x = +3$. This is consistent with the quick analysis of the slope after Figure 2.9.

Likewise, Eq. 2.22 tells you that the slope of z in the y-direction at $x = -3$ is constant and positive, also consistent with the behavior expected from a quick analysis of the shape of the function z.

And just as you can take "higher order" ordinary derivatives such as $\frac{d}{dx}(\frac{dz}{dx}) = \frac{d^2z}{dx^2}$ and $\frac{d}{dy}(\frac{dz}{dy}) = \frac{d^2z}{dy^2}$, you can also take higher-order partial derivatives. So for example $\frac{\partial}{\partial x}(\frac{\partial z}{\partial x}) = \frac{\partial^2 z}{\partial x^2}$ tells you the change in the x-direction slope of z as you move along the x-direction, and $\frac{\partial}{\partial y}(\frac{\partial z}{\partial y}) = \frac{\partial^2 z}{\partial y^2}$ tells you the change in the y-direction slope as you move along the y-direction.

It's important for you to realize that an expression such as $\frac{\partial^2 z}{\partial x^2}$ is the derivative of a derivative, which is *not the same* as $(\frac{\partial z}{\partial x})^2$, which is the square of a first derivative. That's easy to verify for the example given above, in which $\frac{\partial z}{\partial x} = 12xy + 3 + 5y$. In that case, $\frac{\partial^2 z}{\partial x^2} = 12y$, whereas $(\frac{\partial z}{\partial x})^2 = (12xy + 3 + 5y)^2$. By convention the order of the derivative is always written between the "d" or "∂" and the function, as d^2z or $\partial^2 z$, so be sure to look carefully at the location of superscripts when you're dealing with derivatives.

You may also have occasion to use "mixed" partial derivatives such as $\frac{\partial}{\partial x}(\frac{\partial z}{\partial y}) = \frac{\partial^2 z}{\partial x \partial y}$. If you've been tracking the discussion of partial derivatives as slopes of functions in various directions, you can probably guess that $\frac{\partial^2 z}{\partial x \partial y}$ represents the change in the y-direction slope as you move along the x-direction, and $\frac{\partial^2 z}{\partial y \partial x}$ represents the change in the x-direction slope as you move along the y-direction. Thankfully, for well-behaved[5] functions these expressions are

[5] What exactly is a "well-behaved" function? Typically this means any function that is continuous and has continuous derivatives over the region of interest.

interchangeable, so you can take the partial derivatives in either order. You can easily verify this for the example given above by comparing $\frac{\partial}{\partial y}$ of Eq. 2.21 with $\frac{\partial}{\partial x}$ of Eq. 2.22 (the result is $12x + 5$ in both cases).

There's another widely used aspect of partial derivatives you should make sure you understand, and that's the chain rule. Up to this point, we've been dealing with functions such as $z = f(x, y)$ without considering the fact that the variables x and y may themselves be functions of other variables. It's common to call these other variables u and v and to allow both x and y to depend on one or both of u and v. You may encounter situations in which you know the variation in u and v, and you want to know how much your function z will change due to those changes. In such cases, the chain rule for partial derivatives gives you the answer:

$$\frac{\partial z}{\partial u} = \frac{\partial z}{\partial x}\frac{\partial x}{\partial u} + \frac{\partial z}{\partial y}\frac{\partial y}{\partial u}, \tag{2.25}$$

and

$$\frac{\partial z}{\partial v} = \frac{\partial z}{\partial x}\frac{\partial x}{\partial v} + \frac{\partial z}{\partial y}\frac{\partial y}{\partial v}. \tag{2.26}$$

The chain rule is a concise expression of the fact that z depends on both x and y, and since both x and y may change if u changes, the change in z with respect to u is the sum of two terms. The first term is the change in x due to the change in u ($\frac{\partial x}{\partial u}$) times the change in z due to that change in x ($\frac{\partial z}{\partial x}$), and the second term is the change in y due to the change in u ($\frac{\partial y}{\partial u}$) times the change in z due to that change in y ($\frac{\partial z}{\partial y}$). Adding those two terms together gives you Eq. 2.25, and the same reasoning applied to changes in z caused by changes in v leads to Eq. 2.26.

2.6 Vectors as derivatives

In many texts dealing with vectors and tensors, you'll find that vectors are equated to "directional derivatives" and that partial derivatives such as $\frac{\partial}{\partial x}$ and $\frac{\partial}{\partial y}$ are referred to as basis vectors along the coordinate axes.

To understand this correspondence between vectors and derivatives, consider a path such as that shown in Figure 2.10. You can think of this as a path along which you're travelling with velocity $\vec{v}$; for simplicity imagine that this path lies in the xy plane. Now imagine that you're keeping track of time as you move, so you assign a value (such as the t values shown in the figure) to each point on the curve. By marking the curve with values, you have

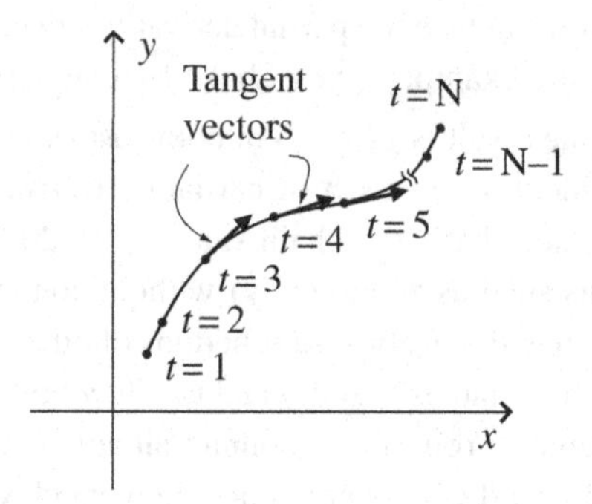

Figure 2.10 Parameterized curve and tangent vectors.

"parameterized" the curve (with t as your parameter).[6] Note that there need not be equal distance along the curve between your parameter values (there definitely won't be if you choose time as your parameter and then change your speed as you move; the reckless driver depicted in Figure 2.10 has apparently sped up in the turn).

As a final bit of visualization, imagine that this curve lies in a region in which the air temperature is different at each location. So as you move along the curve, you will experience the spatial change in air temperature as a temporal change (in other words, you'll be able to make a graph of air temperature vs. time). Of course, how fast the air temperature changes for you will depend both on the distance between measurable changes in the temperature in the direction you're heading and on your speed (how fast you're covering that distance).

With this scenario in mind, the concept of a directional derivative is easy to understand. If the function $f(x, y)$ describes the temperature at each x, y location, the directional derivative ($\frac{df}{dt}$) tells you how much the value of the function f changes as you move a small distance along the curve (in time dt). But recall the chain rule:

$$\frac{df}{dt} = \frac{dx}{dt}\frac{\partial f}{\partial x} + \frac{dy}{dt}\frac{\partial f}{\partial y}. \tag{2.27}$$

This equation says simply that the directional derivative of the function f along the curve parameterized by t (that is, $\frac{df}{dt}$) equals the rate of change of the x-coordinate ($\frac{dx}{dt}$) as you move along the curve times the rate of change of the temperature function with x ($\frac{\partial f}{\partial x}$) plus the rate of change of the y-coordinate ($\frac{dy}{dt}$) as you move along the curve times the rate of change of the temperature

[6] Some authors are careful to distinguish between a "path" and a "curve," using "curve" only when a parameter has been assigned to each point on a path.

function with y $(\frac{\partial f}{\partial y})$. But $(\frac{dx}{dt})$ is just v_x, the x-component of your velocity, and $(\frac{dy}{dt})$ is v_y, the y-component of your velocity. And since you know that your velocity is a vector that is always tangent to the path on which you're moving, you can consider the directional derivative $\frac{df}{dt}$ to be a vector with direction tangent to the curve and with length equal to the rate of change of f with t (that is, the time rate of change of the air temperature).

Now here's the important concept: since f can be any function, you can write Eq. 2.27 as an "operator" equation (that is, an equation waiting to be fed a function on which it can operate):

$$\frac{d}{dt} = \frac{dx}{dt}\frac{\partial}{\partial x} + \frac{dy}{dt}\frac{\partial}{\partial y}. \tag{2.28}$$

The trick to seeing the connection between derivatives and vectors is to view this equation as a vector equation in which

Vector = x-component $\cdot$ x basis vector + y-component $\cdot$ y basis vector.

Comparing this to Eq. 2.28, you should be able to see that the directional derivative operator $\frac{d}{dt}$ represents the tangent vector to the curve, the $\frac{dx}{dt}$ and $\frac{dy}{dt}$ terms represent the x- and y-components of that vector, and the operators $\frac{\partial}{\partial x}$ and $\frac{\partial}{\partial y}$ represent the basis vectors in the direction of the x and y coordinate axes.

Of course, it's not just air temperature that can be represented by $f(x, y)$; this function can represent anything that is spatially distributed in the region around your curve. So $f(x, y)$ could represent the height of the road, the quality of the scenery, or any other quantity that varies in the vicinity of your curve. Likewise, you could have chosen to parameterize your path with markers other than time; had you assigned a value s or λ to each point on your path, the directional derivative $\frac{d}{ds}$ or $\frac{d}{d\lambda}$ would still represent the tangent vector to the curve, $\frac{dx}{ds}$ or $\frac{dx}{d\lambda}$ would still represent the x-component of that vector, and $\frac{dy}{ds}$ or $\frac{dy}{d\lambda}$ would still represent the y-component of that vector.

If you plan to proceed on to the study of tensors, you will find that understanding this relationship between basis vectors along the coordinate axes and partial derivatives is of significant value.

2.7 Nabla – the del operator

The partial derivatives discussed in the previous section can be put to use in a wide range of problems, and when you come across such problems you may find that they involve equations that contain an inverted upper case delta wearing a vector hat ($\vec{\nabla}$). This symbol represents a vector differential operator called "nabla" or "del," and its presence instructs you to take derivatives of the quantity on which the operator is acting. The exact form of those derivatives depends on the symbol following the del operator, with $\vec{\nabla}()$ signifying gradient, "$\vec{\nabla}\circ$" signifying divergence, "$\vec{\nabla}\times$" indicating curl, and $\nabla^2()$ signifying the Laplacian. Each of these operations is discussed in later sections; for now we'll just consider what an operator is and how the del operator can be written in Cartesian coordinates.

Like all good mathematical operators, del is an action waiting to happen. Just as $\sqrt{\ }$ tells you to take the square root of anything that appears under its roof, $\vec{\nabla}$ is an instruction to take derivatives in three directions. Specifically, in Cartesian coordinates

$$\vec{\nabla} \equiv \hat{\imath}\frac{\partial}{\partial x} + \hat{\jmath}\frac{\partial}{\partial y} + \hat{k}\frac{\partial}{\partial z}, \tag{2.29}$$

where $\hat{\imath}$, $\hat{\jmath}$, and $\hat{k}$ are the unit vectors in the direction of the Cartesian coordinates x, y, and z.

This expression may appear strange, since in this form it's lacking anything on which it can operate. However, if you follow the del with a scalar or vector field, you can extract information about how those fields change in space. In this context, "field" refers to an array or collection of values defined at various locations. A scalar field is specified entirely by its magnitude at these locations: examples of scalar fields include the air temperature in a room and the height of terrain above sea level. A vector field is specified by both magnitude and direction at various locations: examples include electric, magnetic, and gravitational fields. Specific examples of how the del operator works on scalar and vector fields are given in the following sections.

2.8 Gradient

When the del operator $\vec{\nabla}$ is followed by a scalar field, the result of the operation is called the gradient of the field. What does the gradient tell you about a scalar field? Two important things: the magnitude of the gradient indicates how quickly the field is changing over space, and the direction of the gradient indicates the direction in which the field is increasing most quickly with distance. So although the gradient operates on a scalar field, the result of the gradient operation is a vector, with both magnitude and direction. Thus, if the scalar field represents terrain height, the magnitude of the gradient at any location tells you how steeply the ground is sloped at that location, and the direction of the gradient points *uphill* along the steepest slope.

The definition of the gradient of the scalar field ψ in Cartesian coordinates is

$$\text{grad}(\psi) = \vec{\nabla}\psi \equiv \hat{\imath}\frac{\partial \psi}{\partial x} + \hat{\jmath}\frac{\partial \psi}{\partial y} + \hat{k}\frac{\partial \psi}{\partial z} \quad \text{(Cartesian).} \tag{2.30}$$

Thus the x-component of the gradient of ψ indicates the slope of the scalar field in the x-direction and the other components indicate the slope in the other directions. The square root of the sum of the squares of these components provides the total steepness of the slope at the location at which the gradient is taken.

You can see a simple example of the result of the gradient operator by considering the tilted plane in Figure 2.11(a). This plane is defined by the simple equation $\psi(x, y) = 5x + 2y$, and you can find the gradient using the two-dimensional version of Eq. 2.30:

$$\begin{aligned}\vec{\nabla}\psi &= \hat{\imath}\frac{\partial(5x + 2y)}{\partial x} + \hat{\jmath}\frac{\partial(5x + 2y)}{\partial y} \\ &= 5\hat{\imath} + 2\hat{\jmath}.\end{aligned}$$

So even though ψ is a scalar function, its gradient is a vector; it has a component along the x-axis and a component along the y-axis. And what do these components tell you?

For one thing, the fact that the x-component is more than twice the size of the y-component tells you that the tilt of the plane is steeper in the x-direction than in the y-direction. You can also tell that the slope in each direction is constant, because the components are not functions of x or y. Both of those conclusions are consistent with Figure 2.11(a).

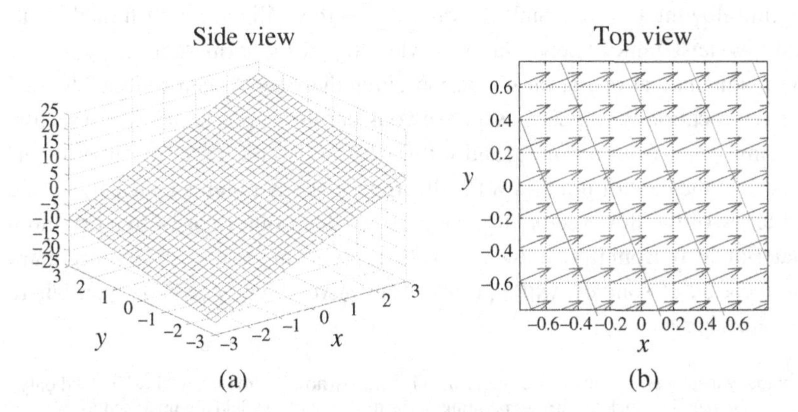

Figure 2.11 Function $\psi = 5x + 2y$ and the gradient and contours of ψ.

And if you wish to determine the magnitude of the gradient, that's easily done. Since the x-component of the gradient is 5 and the y-component is 2, the magnitude of the gradient is simply $(5^2 + 2^2)^{1/2} = 5.39$ over the entire plane. You can also find the angle that the gradient vectors make with the positive x-axis using $\arctan(2/5) = 21.8°$. The gradient and contours of the central portion of the function ψ are shown in Figure 2.11(b).

In cylindrical and spherical coordinates, the gradient is:

$$\vec{\nabla}\psi \equiv \hat{r}\frac{\partial \psi}{\partial r} + \hat{\varphi}\frac{1}{r}\frac{\partial \psi}{\partial \varphi} + \hat{z}\frac{\partial \psi}{\partial z} \quad \text{(cylindrical)}, \tag{2.31}$$

and

$$\vec{\nabla}\psi \equiv \hat{r}\frac{\partial \psi}{\partial r} + \hat{\theta}\frac{1}{r}\frac{\partial \psi}{\partial \theta} + \hat{\varphi}\frac{1}{r\sin\theta}\frac{\partial \psi}{\partial \varphi} \quad \text{(spherical)}. \tag{2.32}$$

You'll see more gradients in Section 2.11 covering the Laplacian operator, which represents the divergence of the gradient. You can read about the divergence in the next section.

2.9 Divergence

When dealing with vector fields, you may encounter the del operator followed by a dot ($\vec{\nabla}\circ$), signifying the divergence of a vector field. The concept of divergence often arises in the areas of physics and engineering that deal with the spatial variation of vector fields, because divergence describes the tendency of vectors to "flow" into or out of a point of interest.[7] Electrostatic fields, for example, may be represented by vectors that point radially away from points at which positive electric charge exists, just as the flow vectors of a fluid point away from a source (such as an underwater spring). Likewise, electrostatic field vectors point toward locations at which negative charge is present, analogous to fluid flowing toward a sink or drain. It was the brilliant Scottish mathematical physicist James Clerk Maxwell who coined the term "convergence" for the mathematical operation which measures the rate of vector "flow" toward a given location. In modern usage we consider the opposite behavior (vectors flowing away from a point), and outward flow is considered positive divergence. In the case of fluid flow, the divergence at any point is a measure of the tendency of the flow vectors to diverge from that point (that is, to carry more material away from it than toward it). Thus points of positive divergence mark the location of sources, while points of negative divergence show you where the sinks are located.

[7] In many instances, nothing in the vector field is actually flowing; the word "flow" is used only as an analogy in which the arrows pointing in the direction of the field are imagined to represent the physical flow of an incompressible fluid.

To understand how this works, take a look at the vector fields shown in Figures 2.12 and 2.13. To find the locations of positive divergence in each of these fields, look for points at which the flow vectors either spread out or are larger pointing away from the location and shorter pointing toward it. Some authors suggest that you imagine sprinkling sawdust on flowing water to assess the divergence; if the sawdust is dispersed, you have selected a point of positive divergence, while if it becomes more concentrated, you've picked a location of negative divergence.

Using such tests, it's clear that locations such as 1 and 2 in Figure 2.12 and locations 4 and 5 in Figure 2.13(a) are points of positive divergence (flow away from these points exceeds flow toward), while the divergence is negative at point 3 in Figure 2.12 (flow toward exceeds flow away).

The divergence at various points in Figure 2.13(b) is less obvious. Location 6 is obviously a point of positive divergence, but what about locations 7 and 8? The flow lines are clearly spreading out at those locations, as they do at location 5 in Figure 2.13(a), but they're also getting shorter pointing away. Does the spreading out compensate for the slowing down of the flow?

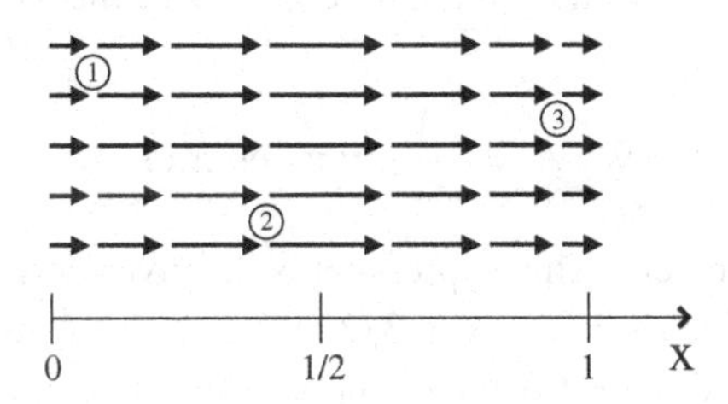

Figure 2.12 Parallel vector field with varying amplitude.

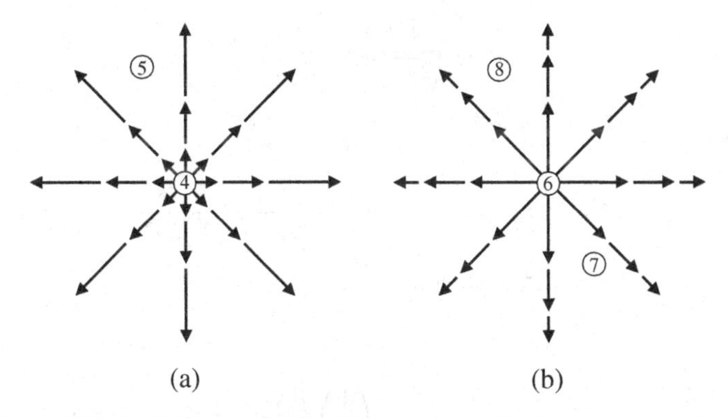

Figure 2.13 Radial vector fields with varying amplitudes.

Answering that question requires a useful mathematical form of the divergence as well as a description of how the vector field varies from place to place. The differential form of the mathematical operation of divergence or "del dot" ($\vec{\nabla}\circ$) on a vector $\vec{A}$ in Cartesian coordinates is

$$\vec{\nabla} \circ \vec{A} = \left(\hat{\imath}\frac{\partial}{\partial x} + \hat{\jmath}\frac{\partial}{\partial y} + \hat{k}\frac{\partial}{\partial z}\right) \circ \left(\hat{\imath}\vec{A}_x + \hat{\jmath}\vec{A}_y + \hat{k}\vec{A}_z\right), \tag{2.33}$$

and, since $\hat{\imath} \circ \hat{\imath} = \hat{\jmath} \circ \hat{\jmath} = \hat{k} \circ \hat{k} = 1$, this is

$$\vec{\nabla} \circ \vec{A} = \left(\frac{\partial A_x}{\partial x} + \frac{\partial A_y}{\partial y} + \frac{\partial A_z}{\partial z}\right). \tag{2.34}$$

Thus the divergence of $\vec{A}$ is simply the change in its x-component along the x-axis plus the change in its y-component along the y-axis plus the change in its z-component along the z-axis. Notice that the divergence of a vector field is a scalar quantity; it has magnitude but no direction.

You can now apply this to the vector field in Figure 2.12. In Figure 2.12, assume that the magnitude of the vector field varies sinusoidally along the x-axis as $\vec{A} = \sin(\pi x)\hat{\imath}$ while remaining constant in the y- and z-directions. Thus,

$$\vec{\nabla} \circ \vec{A} = \frac{\partial A_x}{\partial x} = \pi \cos(\pi x), \tag{2.35}$$

since A_y and A_z are zero. This expression is positive for $0 < x < 1/2$, 0 at $x = 1/2$, and negative for $1/2 < x < 3/2$, just as a visual inspection suggests.

Now consider Figure 2.13(a), which represents a slice through a spherically symmetric vector field with amplitude increasing as the square of the distance from the origin. Thus $\vec{A} = r^2\hat{r}$. Since $r^2 = (x^2 + y^2 + z^2)$ and

$$\hat{r} = \frac{x\hat{\imath} + y\hat{\jmath} + z\hat{k}}{\sqrt{x^2 + y^2 + z^2}},$$

this means

$$\begin{aligned}\vec{A} = r^2\hat{r} &= (x^2 + y^2 + z^2)\frac{x\hat{\imath} + y\hat{\jmath} + z\hat{k}}{\sqrt{x^2 + y^2 + z^2}}\\ &= (x^2 + y^2 + z^2)^{1/2}(x\hat{\imath} + y\hat{\jmath} + z\hat{k}),\end{aligned}$$

and

$$\frac{\partial A_x}{\partial x} = (x^2 + y^2 + z^2)^{1/2} + x\left(\frac{1}{2}\right)(x^2 + y^2 + z^2)^{-1/2}(2x).$$

Doing likewise for the y- and z-components and adding yields

$$\vec{\nabla} \circ \vec{A} = 3(x^2 + y^2 + z^2)^{1/2} + \frac{(x^2 + y^2 + z^2)}{\sqrt{x^2 + y^2 + z^2}} = 4(x^2 + y^2 + z^2)^{1/2} = 4r.$$

Thus the divergence in the vector field in Figure 2.13(a) is increasing linearly with distance from the origin.

Finally, consider the vector field in Figure 2.13(b), which is similar to the previous case but with the amplitude of the vector field *decreasing* as the square of the distance from the origin. The flow lines are spreading out as they were in Figure 2.13(a), but in this case you might suspect that the decreasing amplitude of the vector field will affect the value of the divergence. Since $\vec{A} = (1/r^2)\hat{r}$,

$$\vec{A} = \frac{1}{(x^2 + y^2 + z^2)} \frac{x\hat{\imath} + y\hat{\jmath} + z\hat{k}}{\sqrt{x^2 + y^2 + z^2}} = \frac{x\hat{\imath} + y\hat{\jmath} + z\hat{k}}{(x^2 + y^2 + z^2)^{(3/2)}},$$

and

$$\frac{\partial A_x}{\partial x} = \frac{1}{(x^2 + y^2 + z^2)^{3/2}} - x\left(\frac{3}{2}\right)(x^2 + y^2 + z^2)^{-5/2}(2x).$$

Adding in the y- and z-derivatives gives

$$\vec{\nabla} \circ \vec{A} = \frac{3}{(x^2 + y^2 + z^2)^{3/2}} - \frac{3(x^2 + y^2 + z^2)}{(x^2 + y^2 + z^2)^{5/2}} = 0.$$

This validates the suspicion that the reduced amplitude of the vector field with distance from the origin may compensate for the spreading out of the flow lines. Note that this is true only for the case in which the amplitude of the vector field falls off as $1/r^2$ (and only for points away from the origin).[8] Therefore, you must consider two key factors in determining the divergence at any point: the *spacing* and the *relative amplitudes* of the field lines at that point. These factors both contribute to the total flow of field lines into or out of an infinitesimally small volume around the point. If the outward flow exceeds the inward flow, the divergence is positive at that point. If the outward flow is less than the inward flow, the divergence is negative, and if the outward and inward flows are equal the divergence is zero at that point.

So far the divergence has been calculated for the Cartesian coordinate system, but depending on the symmetries of the problem, it might be solved

[8] At the origin, where $r = 0$, a $(1/r^2)$-vector field experiences a singularity, and the Dirac delta function must be employed to determine the divergence.

more easily using non-Cartesian systems. The divergence may be calculated in cylindrical and spherical coordinate systems using

$$\vec{\nabla} \circ \vec{A} = \frac{1}{r}\frac{\partial}{\partial r}(rA_r) + \frac{1}{r}\frac{\partial A_\phi}{\partial \phi} + \frac{\partial A_z}{\partial z}, \quad \text{(cylindrical)} \tag{2.36}$$

and

$$\vec{\nabla} \circ \vec{A} = \frac{1}{r^2}\frac{\partial}{\partial r}(r^2 A_r) + \frac{1}{r \sin\theta}\frac{\partial}{\partial \theta}(A_\theta \sin\theta) + \frac{1}{r \sin\theta}\frac{\partial A_\phi}{\partial \phi}. \quad \text{(spherical)} \tag{2.37}$$

If you doubt the efficacy of choosing the proper coordinate system, you should re-work the last two examples in this section using spherical coordinates.

2.10 Curl

The del operator followed by a cross ($\vec{\nabla}\times$) signifies the differential operation of curl. The curl of a vector field is a measure of the field's tendency to circulate about a point, much like the divergence is a measure of the tendency of the field to flow away from a point. But unlike the divergence, which produces a scalar result, the curl produces a vector. The magnitude of the curl vector is proportional to the amount of circulation of the field around the point of interest, and the direction of the curl vector is perpendicular to the plane in which the field's circulation is a maximum.

The curl at a point in a vector field can be understood by considering the vector fields shown in Figure 2.14. To find the locations of large curl in each of these fields, look for points at which the flow vectors on one side of the point are significantly different (in magnitude, direction, or both) from the

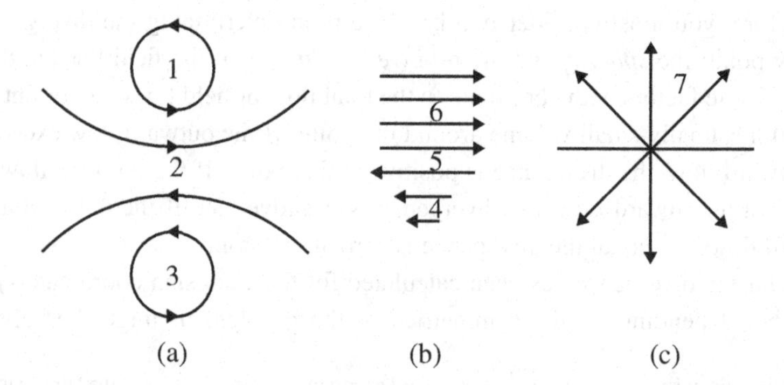

Figure 2.14 Vector fields with various values of curl.

flow vectors on the opposite side of the point. Once again a thought experiment is helpful: imagine holding a tiny paddlewheel at each point in the flow. If the flow would cause the paddlewheel to rotate, the center of the wheel marks a point of non-zero curl. The direction of the curl is along the axis of the paddlewheel. By convention, the positive-curl direction is determined by the right-hand rule: if you curl the fingers of your right hand along the circulation direction, your thumb points in the direction of positive curl.

Using the paddlewheel test, you can see that points 1, 2, and 3 in Figure 2.14(a) and point 5 in Figure 2.14(b) are high-curl locations, and some curl also exists at point 4. The uniform flow around point 6 and the diverging flow lines around Point 7 in Figure 2.14(c) would not cause a tiny paddlewheel to rotate, meaning that these are points of low or zero curl.

To make this quantitative, you can use the differential form of the curl or "del cross" ($\vec{\nabla}\times$) operator in Cartesian coordinates:

$$\vec{\nabla} \times \vec{A} = \left(\hat{\imath}\frac{\partial}{\partial x} + \hat{\jmath}\frac{\partial}{\partial y} + \hat{k}\frac{\partial}{\partial z}\right) \times \left(\hat{\imath}A_x + \hat{\jmath}A_y + \hat{k}A_z\right). \tag{2.38}$$

Recall that the vector cross-product may be written as a determinant:

$$\vec{\nabla} \times \vec{A} = \begin{vmatrix} \hat{\imath} & \hat{\jmath} & \hat{k} \\ \frac{\partial}{\partial x} & \frac{\partial}{\partial y} & \frac{\partial}{\partial z} \\ A_x & A_y & A_z \end{vmatrix}, \tag{2.39}$$

which expands to

$$\vec{\nabla} \times \vec{A} = \left(\frac{\partial A_z}{\partial y} - \frac{\partial A_y}{\partial z}\right)\hat{\imath} + \left(\frac{\partial A_x}{\partial z} - \frac{\partial A_z}{\partial x}\right)\hat{\jmath} + \left(\frac{\partial A_y}{\partial x} - \frac{\partial A_x}{\partial y}\right)\hat{k}. \tag{2.40}$$

Notice that each component of the curl of $\vec{A}$ indicates the tendency of the field to rotate in one of the coordinate planes. If the curl of the field has a large x-component, it means that the field has significant circulation about that point in the yz plane. The overall direction of the curl represents the axis about which the rotation is greatest, with the sense of the rotation given by the right-hand rule.

If you're wondering how the terms in this equation measure rotation, consider the vector fields shown in Figure 2.15. Look first at the field in Figure 2.15(a) and the x-component of the curl in the equation: this term involves the change in A_z with y and the change in A_y with z. Proceeding in the positive y-direction from the left side of the point of interest to the right, A_z is clearly increasing (it's pointing in the negative z-direction on the left side of the point of interest and the positive z-direction on the right side), so the term $\frac{\partial A_z}{\partial y}$ must be positive. Looking now at A_y, you can see that it is positive below

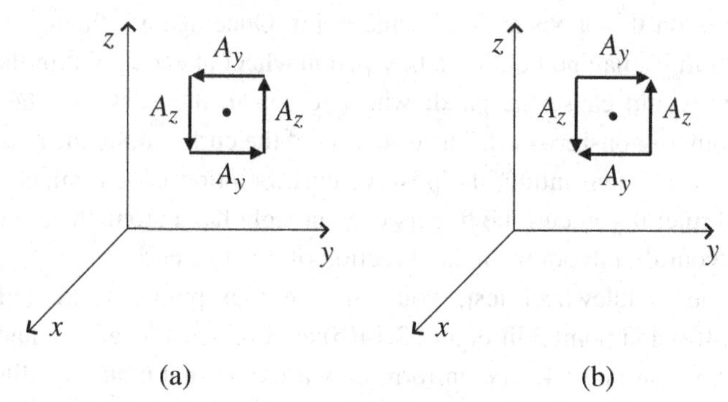

Figure 2.15 Effect of $\frac{\partial A_y}{\partial z}$ and $\frac{\partial A_z}{\partial y}$ on the value of the curl.

the point of interest and negative above, so it is decreasing in the positive z-direction. Thus $\frac{\partial A_y}{\partial z}$ is negative, which means that it increases the value of the curl when it is subtracted from $\frac{\partial A_z}{\partial y}$. Thus the curl has a large value at the point of interest, as expected in light of the circulation of $\vec{A}$ about this point.

The situation in Figure 2.15(b) is quite different. In this case, both $\frac{\partial A_y}{\partial z}$ and $\frac{\partial A_z}{\partial y}$ are positive, and subtracting $\frac{\partial A_y}{\partial z}$ from $\frac{\partial A_z}{\partial y}$ gives a small result. The value of the x-component of the curl is therefore small in this case. Vector fields with zero curl at all points are called "irrotational."

Here are expressions for the curl in cylindrical and spherical coordinates:

$$\vec{\nabla}\times\vec{A} = \left(\frac{1}{r}\frac{\partial A_z}{\partial \phi} - \frac{\partial A_\phi}{\partial z}\right)\hat{r} + \left(\frac{\partial A_r}{\partial z} - \frac{\partial A_z}{\partial r}\right)\hat{\phi} + \frac{1}{r}\left(\frac{\partial (rA_\phi)}{\partial r} - \frac{\partial A_r}{\partial \phi}\right)\hat{z}, \quad \text{(cylindrical)} \tag{2.41}$$

$$\vec{\nabla}\times\vec{A} = \frac{1}{r\sin\theta}\left(\frac{\partial (A_\phi \sin\theta)}{\partial \theta} - \frac{\partial A_\theta}{\partial \phi}\right)\hat{r} + \frac{1}{r}\left(\frac{1}{\sin\theta}\frac{\partial A_r}{\partial \phi} - \frac{\partial (rA_\phi)}{\partial r}\right)\hat{\theta} + \frac{1}{r}\left(\frac{\partial (rA_\theta)}{\partial r} - \frac{\partial A_r}{\partial \theta}\right)\hat{\phi}. \quad \text{(spherical)} \tag{2.42}$$

A common misconception is that the curl of a vector field is non-zero wherever the field appears to curve. However, just as the divergence depended both on the spreading out and the changing length of field lines, the curl depends not only on the curvature of the lines but also on the strength of the field. Consider a curving field that points in the $\hat{\phi}$ direction and decreases as $1/r$:

$$\vec{A} = \frac{k}{r}\hat{\phi}.$$

Finding the curl of this field is particularly straightforward in cylindrical coordinates:

$$\vec{\nabla} \times \vec{A} = \left(\frac{1}{r}\frac{\partial A_z}{\partial \phi} - \frac{\partial A_\phi}{\partial z}\right)\hat{r} + \left(\frac{\partial A_r}{\partial z} - \frac{\partial A_z}{\partial r}\right)\hat{\phi} + \frac{1}{r}\left(\frac{\partial (rA_\phi)}{\partial r} - \frac{\partial A_r}{\partial \phi}\right)\hat{z}.$$

Since A_r and A_z are both zero, this is

$$\vec{\nabla} \times \vec{A}$$
$$= \left(-\frac{\partial A_\phi}{\partial z}\right)\hat{r} + \frac{1}{r}\left(\frac{\partial (rA_\phi)}{\partial r}\right)\hat{z} = \left(-\frac{\partial (k/r)}{\partial z}\right)\hat{r} + \frac{1}{r}\left(\frac{\partial (rk/r)}{\partial r}\right)\hat{z} = 0.$$

To understand the physical basis for this result, consider again the fluid-flow and paddlewheel analogy. Imagine the forces on the paddlewheel placed in the field shown in Figure 2.16(a). The center of curvature is well below the bottom of the figure, and the spacing of the arrows indicates that the field is getting weaker with distance from the center. At first glance, it may seem that this paddlewheel would rotate clockwise due to the curvature of the field, since the flow lines are pointing slightly upward at the left paddle and slightly downward at the right. But consider the effect of the weakening of the field above the axis of the paddlewheel: the top paddle receives a weaker push from the field than the bottom paddle, as shown in Figure 2.16(b). The stronger force on the bottom paddle will attempt to cause the paddlewheel to rotate counter-clockwise. Thus the downward curvature of the field is offset by the weakening of the field with distance from the center of curvature. And if the field diminishes as $1/r$, the upward-downward push on the left and right paddles is exactly compensated by the weaker-stronger push on the top and bottom paddles. The clockwise and counter-clockwise forces balance, and the paddlewheel does not turn – the curl at this location is zero, even though the field lines are curved.

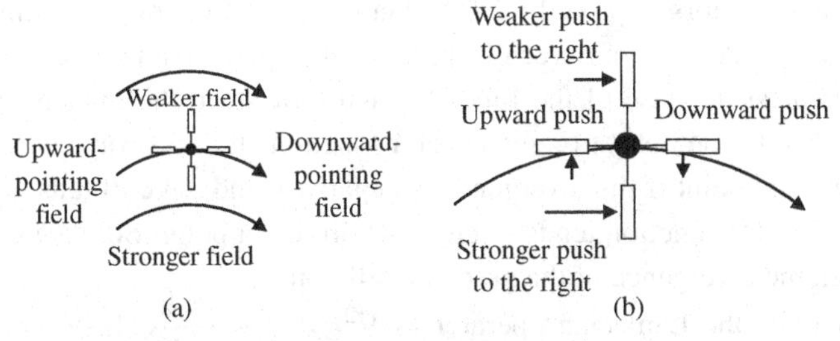

Figure 2.16 Offsetting components of the curl of $\vec{A}$.

For this $1/r$ field, the curl is zero everywhere except at the center of curvature (where a singularity exists and must be handled using the delta function).

2.11 Laplacian

Once you know that the gradient operates on a scalar function and produces a vector and that the divergence operates on a vector and produces a scalar, it's natural to wonder whether these two operations can be combined in a meaningful way. As it turns out, the divergence of the gradient of a scalar function ϕ, written as $\vec{\nabla} \circ (\vec{\nabla}\phi)$, is one of the most useful mathematical operations in physics and engineering. This operation, usually written as $\nabla^2\phi$ (but sometimes as $\Delta\phi$), is called the "Laplacian" in honor of Pierre-Simon Laplace, the great French mathematician and astronomer.

Before trying to understand why the Laplacian operator is so valuable, you should begin by recalling the operations of gradient and divergence in Cartesian coordinates:

Gradient:

$$\vec{\nabla}\phi = \hat{\imath}\frac{\partial \phi}{\partial x} + \hat{\jmath}\frac{\partial \phi}{\partial y} + \hat{k}\frac{\partial \phi}{\partial z}. \tag{2.43}$$

Divergence:

$$\vec{\nabla} \circ \vec{A} = \frac{\partial A_x}{\partial x} + \frac{\partial A_y}{\partial y} + \frac{\partial A_z}{\partial z}. \tag{2.44}$$

Since the x-component of the gradient of ϕ is $\frac{\partial \phi}{\partial x}$, the y-component of the gradient of ϕ is $\frac{\partial \phi}{\partial y}$, and the z-component of the gradient of ϕ is $\frac{\partial \phi}{\partial z}$, the divergence of the vector produced by the gradient is

$$\vec{\nabla} \circ \vec{\nabla}\phi = \nabla^2\phi = \frac{\partial^2 \phi}{\partial x^2} + \frac{\partial^2 \phi}{\partial y^2} + \frac{\partial^2 \phi}{\partial z^2}. \tag{2.45}$$

Just as the gradient ($\vec{\nabla}$), divergence ($\vec{\nabla}\circ$), and curl ($\vec{\nabla}\times$) represent differential operators, so too the Laplacian (∇^2) is an operator waiting to be fed a function. As you may recall, the gradient operator tells you the direction of greatest increase of the function (and how steep the increase is), the divergence tells you how strongly a vector function "flows" away from a point (or toward that point if the divergence is negative), and the curl tells you how strongly a vector function tends to circulate around a point. So what does the Laplacian, the divergence of the gradient, tell you?

If you write the Laplacian operator as $\nabla^2 = \frac{\partial^2}{\partial x^2} + \frac{\partial^2}{\partial y^2} + \frac{\partial^2}{\partial z^2}$, it should help you see that this operator finds the *change in the change* of the function (if you make a graph, the change in the slope) in all directions from the point of interest. That may not seem very interesting, until you consider that acceleration is the change in the change of position with time, or that the maxima and minima of functions (peaks and valleys) are regions in which the slope changes significantly, or that one way to find blobs and edges in a digital image is to look for points at which the gradient of the brightness suddenly changes.

To understand why the Laplacian performs such a diverse set of useful tasks, it helps to understand that at each point in space, the Laplacian of a function represents the difference between the value of the function at that point and the average of the values at surrounding points. How does it do that? Consider the region around the point labeled (0, 0, 0) in Figure 2.17. The function ϕ exists in all three dimensions around this region, and the cube is shown only to illustrate the location of six points around the central point (0, 0, 0), where the value of the function ϕ is ϕ_0. Notice that there are points in front of and behind the central point (along the x-axis), points to the left and right (along the y-axis), and points above and below (along the z-axis). To see how the change in the change in ϕ is related to ϕ_0, consider for now the points along the x-axis, as shown in Figure 2.18. Notice that the value of ϕ at the point in back of the central point is labeled ϕ_{Back} and the value of ϕ in front of the central point is labeled ϕ_{Front}. If each of these points is located a distance of Δx

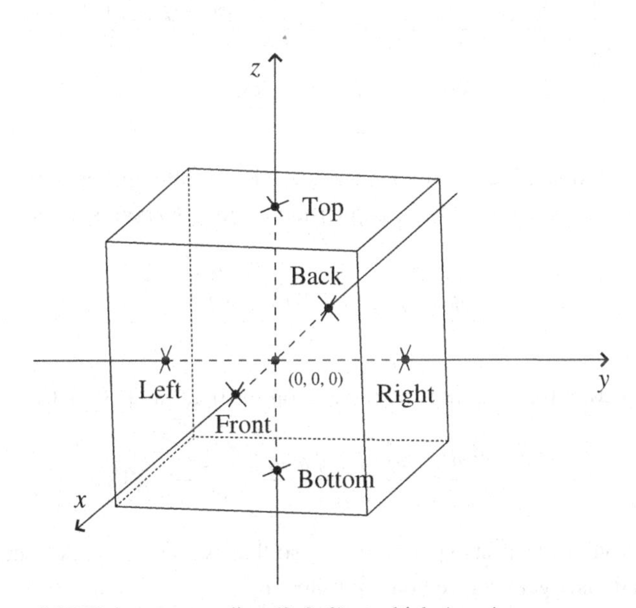

Figure 2.17 Points surrounding (0, 0, 0) at which $\phi = \phi_0$.

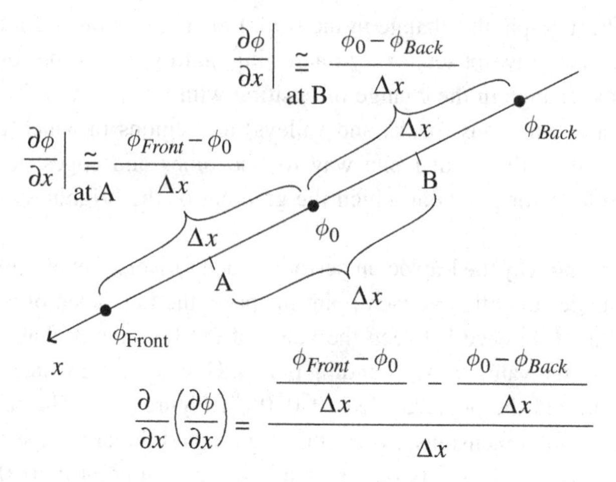

Figure 2.18 Change in ϕ along x-axis.

from (0, 0, 0), then the partial derivative of ϕ at point B can be approximated by $(\phi_0 - \phi_{Back})/\Delta x$. Likewise, the partial derivative of ϕ at point A can be approximated by $(\phi_{Front} - \phi_0)/\Delta x$.

But the Laplacian involves not just the change in ϕ, but the *change in the change* of ϕ. For that, you can write

$$\frac{\partial}{\partial x}\left(\frac{\partial \phi}{\partial x}\right) = \frac{(\phi_{Front} - \phi_0)/\Delta x - (\phi_0 - \phi_{Back})/\Delta x}{\Delta x},$$

$$\frac{\partial^2 \phi}{\partial x^2} = \frac{\phi_{Front} + \phi_{Back} - 2\phi_0}{\Delta x^2}. \tag{2.46}$$

And although this might not look very helpful, good things happen when you combine this expression with the expression for the two points to the right and left of (0, 0, 0):

$$\frac{\partial^2 \phi}{\partial y^2} = \frac{\phi_{Right} + \phi_{Left} - 2\phi_0}{\Delta y^2}, \tag{2.47}$$

and the equation for the points on top and on the bottom of (0, 0, 0):

$$\frac{\partial^2 \phi}{\partial z^2} = \frac{\phi_{Top} + \phi_{Bottom} - 2\phi_0}{\Delta z^2}. \tag{2.48}$$

If you pick your locations symmetrically so that $\Delta x = \Delta y = \Delta z$, then these three equations together give you the following:

$$\frac{\partial^2 \phi}{\partial x^2} + \frac{\partial^2 \phi}{\partial y^2} + \frac{\partial^2 \phi}{\partial z^2}$$
$$= \frac{\phi_{Front} + \phi_{Back} + \phi_{Right} + \phi_{Left} + \phi_{Top} + \phi_{Bottom} - 6\phi_0}{\Delta x^2}. \quad (2.49)$$

Using the del-squared notation for the Laplacian and a little rearranging makes this

$$\nabla^2 \phi = \frac{-6}{\Delta x^2}\left[\phi_0 - \frac{1}{6}(\phi_{Front} + \phi_{Back} + \phi_{Right} + \phi_{Left} + \phi_{Top} + \phi_{Bottom})\right]$$
$$= \frac{-6}{\Delta x^2}(\phi_0 - \phi_{avg}), \quad (2.50)$$

where the average value of the function ϕ over the six surrounding points is $\phi_{avg} = \frac{1}{6}(\phi_{Front} + \phi_{Back} + \phi_{Right} + \phi_{Left} + \phi_{Top} + \phi_{Bottom})$.

Equation 2.50 tells you that the Laplacian of a function ϕ at any point is proportional to the difference between the value of ϕ at that point and the average value of ϕ at the surrounding points. The negative sign in this equation tells you that the Laplacian is negative if the value of the function at the point of interest is greater than the average of the function's value at the surrounding points, and the Laplacian is positive if the value at the point of interest is smaller than the average of the value at the surrounding points.

And how does the difference between a function's value at a point and the average value at neighboring points relate to the divergence of the gradient of that function? To understand that, think about a point at which the function's value is greater than the surrounding average – such a point represents a local maximum of the function. Likewise, a point at which the function's value is less than the surrounding average represents a local minimum. This is the reason you may find the Laplacian described as a "concavity detector" or a "peak finder" – it finds points at which the value of the function sticks above or falls below the values at the surrounding points.

To better understand how peaks and valleys relate to the divergence of the gradient of a function, recall that the gradient points in direction of steepest incline (or decline if the gradient is negative), and divergence measures the "flow" of a vector field out of a region (or into the region if the divergence is negative). Now consider the peak of the function shown in Figure 2.19(a) and the gradient of the function in the vicinity of that peak, shown in Figure 2.19(b). Near the peak, the gradient vectors "flow" toward the peak from all directions. Vector fields that converge upon a point have negative divergence, so this means that the divergence of the gradient in the vicinity of a

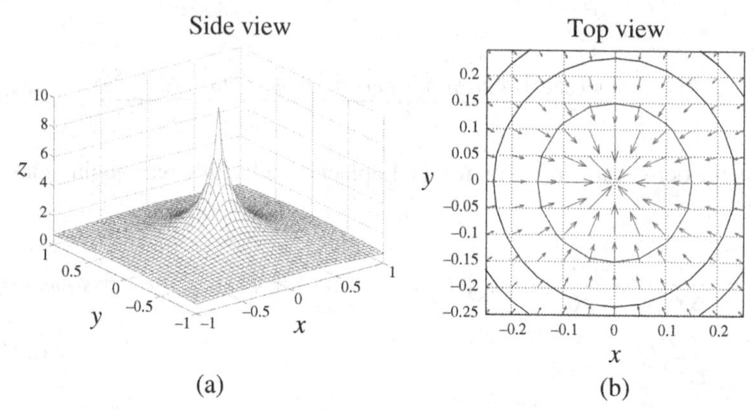

Figure 2.19 Function ϕ (varying as 1/r) and the gradient and contours of ϕ near the peak.

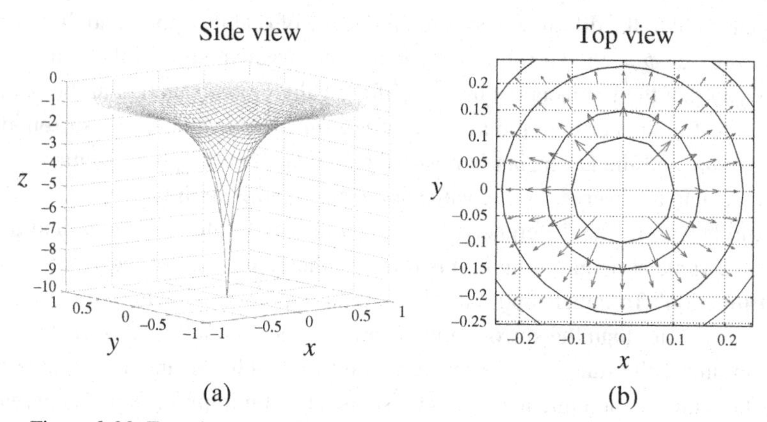

Figure 2.20 Function ϕ (varying as $-1/r$) and the gradient and contours of ϕ near the bottom of the valley.

peak will be a large negative number. This is consistent with the conclusion that the Laplacian is negative near a function's maximum point.

The alternative case is shown in Figures 2.20(a) and 2.20(b). Near the bottom of a valley, the gradient "flows" outward in all directions, so the divergence of the gradient is a large positive number in this case (again consistent with the conclusion that the Laplacian of a minimum point is positive). And what is the value of the Laplacian of a function away from a peak or valley? The answer to that question depends on the shape of the function in the vicinity of the point in question. As described in Section 2.9, the value of the divergence depends on how strongly the function "flows" away from a small volume surrounding the point of interest. Since the Laplacian involves the divergence of the gradient, the question is whether the gradient vectors "flow" toward or away from the point (in other words, whether the gradient vectors tend to concentrate toward or disperse away from that point). If the inward flow of gradient vectors equals the outward flow, then the Laplacian of the function is zero at that point. But if the length and direction of the gradient vectors conspire to make the outward flow greater than the inward flow at some point, then the Laplacian is positive at that point.

For example, if you're climbing out of a circularly symmetric valley with constant slope, the gradient vectors are spreading apart without changing in length, which means the divergence of the gradient (and hence the Laplacian) will have a positive value at that point. But if a different valley has walls for which the slope gets less steep (so the gradient vectors get shorter) as you move away from the bottom of the valley, it's possible for the reduced strength of the gradient vectors to exactly compensate for the spreading apart of those vectors, in which case the Laplacian will be zero.

To see how this works mathematically, consider a three-dimensional function ϕ whose value decreases in inverse proportion to the distance r from the origin. This function may be written as $\phi = k/r$, where k is just a constant of proportionality and r is the distance from the origin. Thus $r = (x^2+y^2+z^2)^{1/2}$ and $\phi = k/(x^2 + y^2 + z^2)^{1/2}$. You can find the value of the Laplacian for this case using Eq. 2.45; the first step is to find the partial derivative of ϕ with respect to x

$$\frac{\partial \phi}{\partial x} = \frac{-k(2x)}{2(x^2 + y^2 + z^2)^{3/2}} = \frac{-kx}{(x^2 + y^2 + z^2)^{3/2}},$$

after which you take another partial with respect to x:

$$\begin{aligned}\frac{\partial^2 \phi}{\partial x^2} &= \frac{-k}{(x^2 + y^2 + z^2)^{3/2}} + \left(\frac{3}{2}\right)\frac{kx(2x)}{(x^2 + y^2 + z^2)^{5/2}} \\ &= \frac{-k}{(x^2 + y^2 + z^2)^{3/2}} + \frac{3kx^2}{(x^2 + y^2 + z^2)^{5/2}}.\end{aligned}$$

The same approach for the second-order partials with respect to y and z gives

$$\frac{\partial^2 \phi}{\partial y^2} = \frac{-k}{(x^2 + y^2 + z^2)^{3/2}} + \frac{3ky^2}{(x^2 + y^2 + z^2)^{5/2}},$$

and

$$\frac{\partial^2 \phi}{\partial z^2} = \frac{-k}{(x^2 + y^2 + z^2)^{3/2}} + \frac{3kz^2}{(x^2 + y^2 + z^2)^{5/2}}.$$

Now it's just a matter of adding all three second-order partials:

$$\frac{\partial^2\phi}{\partial x^2}+\frac{\partial^2\phi}{\partial y^2}+\frac{\partial^2\phi}{\partial z^2}=\frac{-3k}{(x^2+y^2+z^2)^{3/2}}+\frac{3k(x^2+y^2+z^2)}{(x^2+y^2+z^2)^{5/2}}$$
$$=\frac{-3k}{(x^2+y^2+z^2)^{3/2}}+\frac{3k}{(x^2+y^2+z^2)^{3/2}}=0.$$

So for a three-dimensional function with $1/r$-dependence, the Laplacian of the function is zero everywhere away from the origin. What about at the origin itself? That point requires special treatment, since the $1/r$-dependence of the function becomes problematic at $r = 0$. That special treatment involves the Dirac delta function and integral rather than differential techniques.

You may occasionally have need to calculate the Laplacian in non-Cartesian coordinates. For function ψ, the Laplacian in cylindrical and spherical coordinates is given by:

Cylindrical

$$\nabla^2\psi=\frac{1}{r}\frac{\partial}{\partial r}\left(r\frac{\partial\psi}{\partial r}\right)+\frac{1}{r^2}\frac{\partial^2\psi}{\partial\phi^2}+\frac{\partial^2\psi}{\partial z^2}, \tag{2.51}$$

Spherical

$$\nabla^2\psi=\frac{1}{r^2}\frac{\partial}{\partial r}\left(r^2\frac{\partial\psi}{\partial r}\right)+\frac{1}{r^2\sin\theta}\frac{\partial}{\partial\theta}\left(\sin\theta\frac{\partial\psi}{\partial\theta}\right)+\frac{1}{r^2\sin^2\theta}\frac{\partial^2\psi}{\partial\phi^2}. \tag{2.52}$$

2.12 Chapter 2 problems

2.1 For vectors $\vec{A} = 3\hat{\imath} + 2\hat{\jmath} - \hat{k}$ and $\vec{B} = \hat{\jmath} + 4\hat{k}$, find the scalar product $\vec{A} \circ \vec{B}$ and the angle between $\vec{A}$ and $\vec{B}$.

2.2 If vector $\vec{J} = 2\hat{\imath} - \hat{\jmath} + 5\hat{k}$ and $\vec{K} = 3\hat{\imath} + 2\hat{\jmath} + \hat{k}$, find the vector $\vec{L}$ that equals the cross product $\vec{J} \times \vec{K}$. Also show that $\vec{L}$ is perpendicular to both $\vec{J}$ and to $\vec{K}$.

2.3 Show that $\vec{A} \circ \vec{B} = A_xB_x + A_yB_y + A_zB_z = |\vec{A}||\vec{B}/\cos(\theta)$ and that $|\vec{A} \times \vec{B}| = |\vec{A}||\vec{B}/\sin(\theta)$.

2.4 Using the vectors of the previous two problems, find the triple product $\vec{J} \circ (\vec{A} \times \vec{B})$. Compare your answer to $(\vec{J} \times \vec{A}) \circ \vec{B}$.

2.5 Using the vectors of Problems 1 and 2, find the triple vector product $\vec{J} \times (\vec{A} \times \vec{B})$. Compare your answer to $(\vec{J} \times \vec{A}) \times \vec{B}$ and to $\vec{B} \times (\vec{J} \times \vec{A})$.

2.6 For the function $f(x, y) = x^2 + 3y^2 + 2xy + 3x + 5$, find $\frac{\partial f}{\partial x}$ and $\frac{\partial f}{\partial y}$.

2.7 If $\phi = x^2 + y^2$, what is $\vec{\nabla}\phi$ at the position $(x, y) = (3\text{ cm}, -2\text{ cm})$?

2.8 Find the divergence of the vector field given by $\vec{C} = 5xy\hat{\imath} - 3x\hat{\jmath} + 5z^2\hat{k}$.

2.9 What is the curl of the vector field given in the previous problem?

2.10 Find the Laplacian of the function given in Problem 2.6.

2.11 In mechanics, the work (W) done by a force ($\vec{F}$) acting over a displacement ($\vec{dr}$) is defined as the scalar product between the force and the displacement, so $W = \vec{F} \circ \vec{dr}$. How much work is done by the vertically downward force of Earth's gravity ($|\vec{F}| = mg$, where g is the acceleration of gravity) on a car with a mass of 1200 kg as the car moves 50 meters down a hill whose surface makes an angle of 20 degrees below the horizontal?

2.12 Imagine trying to turn the head of a bolt by pushing on the handle of a wrench. The vector torque exerted by the force you apply ($\vec{F}$) is given by the equation $\vec{\tau} = \vec{r} \times \vec{F}$, where $\vec{r}$ is a vector from the point of rotation to the point of application of the force. If you push on the handle of the wrench with a force of 25 N at a distance of 12 cm from the point of rotation, in what direction should you push to maximize the torque on the bolt head? If you push in that direction, how much torque will you exert on the bolt head?

3
Vector applications

The real value of understanding vectors and how to manipulate them becomes clear when you realize that your knowledge allows you to solve a variety of problems that would be much more difficult without vectors. In this chapter, you'll find detailed explanations of four such problems: a mass sliding down an inclined plane, an object moving along a curved path, a charged particle in an electric field, and a charged particle in a magnetic field. To solve these problems, you'll need many of the vector concepts and operations described in Chapters 1 and 2.

3.1 Mass on an inclined plane

Consider the delivery woman pushing a heavy box up the ramp to her delivery truck, as illustrated in Figure 3.1. In this situation, there are a number of forces acting on the box, so if you want to determine how the box will move, you need to know how to work with vectors. Specifically, to solve problems such as this, you can use vector addition to find the total force acting on the box, and then you can use Newton's Second Law to relate that total force to the acceleration of the box.

To understand how this works, imagine that the delivery woman slips off the side of the ramp, leaving the box free to slide down the ramp under the influence of gravity. For starters, pretend that the ramp is so slippery that friction between the bottom of the box and the ramp surface is negligible (so the coefficient of friction is effectively zero). How fast will the box be moving when it reaches the bottom of the ramp? Perhaps more importantly, on what does that speed depend?

Whenever you approach a problem like this, it's a good idea to begin by drawing a diagram that shows all the forces acting on the box. Such a

Figure 3.1 The delivery-truck problem.

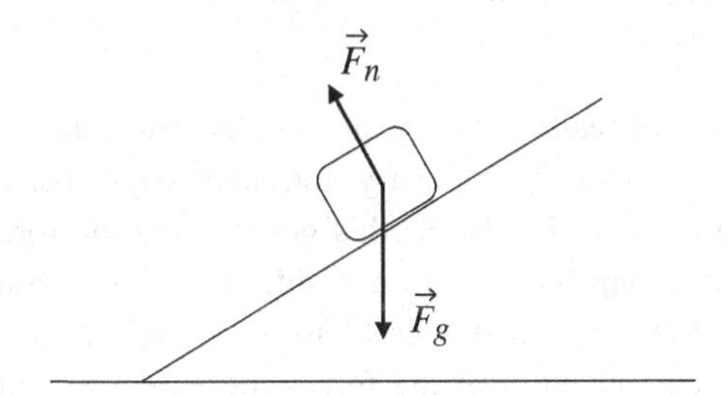

Figure 3.2 Free-body diagram for mass on frictionless ramp.

"free-body" diagram will help you determine the total force acting on the object, from which you can easily determine the object's acceleration using Newton's Second Law ($\vec{a} = \Sigma\vec{F}/m$).[1] And once you know the acceleration, it's an easy matter to find the velocity. An example of the free-body diagram for this (frictionless) case is shown in Figure 3.2.

By removing the delivery woman and friction from the problem, the only remaining forces acting on the box are the force of gravity $\vec{F}_g$, which points vertically downward,[2] and the normal force $\vec{F}_n$, which is perpendicular (or "normal") to the surface of the ramp. The origin of these forces is easy to understand; the gravitational force is produced by the mass of the Earth, and the normal force is produced by the ramp as a reaction to the force produced by the box on the ramp (if the ramp weren't pushing upward on the box, gravity would cause the box to accelerate straight downward).

[1] You may be more accustomed to seeing this as $\vec{F} = m\vec{a}$, but the form shown above is meant to remind you that it's the *sum of the forces* that produces acceleration, and the primary job of all mass is to *resist* acceleration (which is why mass lives in the denominator – if the same force is applied to a large mass and a small mass, the small mass experiences greater acceleration).

[2] This ignores local gravitational anomalies, which is a very reasonable thing to do for problems of this type.

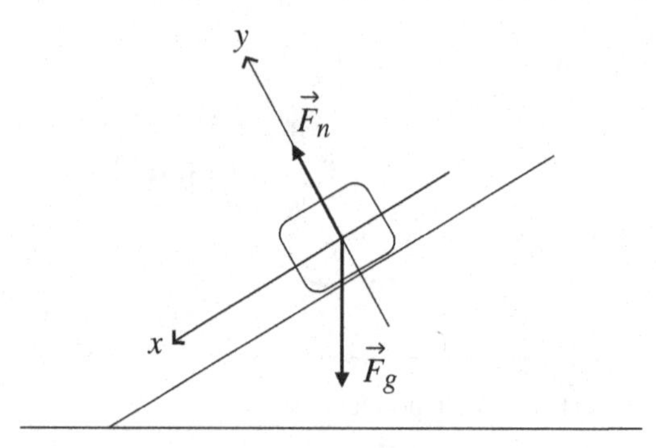

Figure 3.3 Free-body diagram with coordinate axes.

Do these two forces really act only at a single point somewhere inside the box, as implied by Figure 3.2? Clearly not, since every particle in the box is being pulled downward by the Earth's gravity, and the force of the ramp on the box occurs along the entire underside of the box. But to determine the acceleration of the box in this problem, you don't need to worry about the actual point of application of the forces, because you can treat the box as a *particle* that exists at a single location. That's not always the case; in problems involving torque and angular acceleration, for example, the point of application of the force may be critically important. But the box in this problem is sliding, not rolling, down the ramp, and you're perfectly justified in treating the box as a single particle and drawing the forces as though they all act at the same point. Furthermore, you're less likely to make a mistake about the angles of the forces if you draw them as in Figure 3.2. This approach can be justified using the concept of center of mass (CM), since for a rigid object of mass m you can consider the entire object as a single point and write $\vec{a}_{CM} = \vec{F}_{CM}/m$.

Before doing the vector addition of the two forces acting on the box to determine the total force, it's a good idea to draw a set of coordinate axes onto your free-body diagram, as in Figure 3.3. Of course, you're free to draw the axes in any direction you choose, but when you're faced with a problem of a mass on an inclined plane, there are certain benefits to drawing the x-axis pointing down the ramp (and parallel to the ramp surface) and the y-axis pointing upward (and perpendicular to the ramp surface). This approach has the advantage that the normal force lies entirely along the positive y-axis, and the motion of the block sliding down the ramp is entirely in the positive x-direction (as long as the box stays on the ramp). To pay for that advantage, you'll have to use a bit of geometry to find the x- and y-components of the gravitational

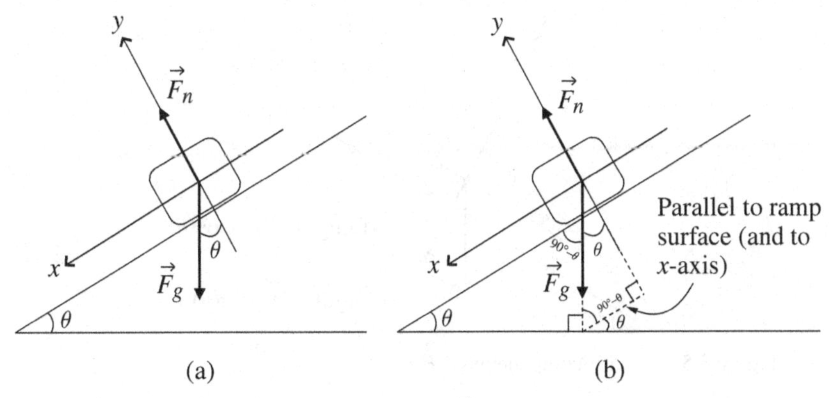

Figure 3.4 Geometry to find the angle of $\vec{F}_g$.

force, since the vector $\vec{F}_g$ points straight downward and is therefore aligned with neither the down-plane (x-) nor the perpendicular-to-plane (y-) axis.[3]

The key to finding the x-component ($\vec{F}_{g,x}$) and the y-component ($\vec{F}_{g,y}$) of the gravitational force ($\vec{F}_g$) is to realize that the angle θ between the ramp surface and the horizontal is also the angle between $\vec{F}_g$ and the negative y-axis, as shown in Figure 3.4(a).

If you're uncertain why the two angles shown as θ in Figure 3.4(a) must be the same, take a look at Figure 3.4(b). Completing the two triangles shown in Figure 3.4(b) should help you see that the angle between $\vec{F}_g$ and the negative y-axis is indeed θ (you may also be able to see this by imagining the case in which $\theta = 0°$ or $\theta = 90°$).

Once you're convinced that the angle between $\vec{F}_g$ and the negative y-axis is θ, it's quite straightforward to determine $\vec{F}_{g,x}$ and $\vec{F}_{g,y}$, the x- and y-components of the gravitational force vector $\vec{F}_g$. As you can see in Figure 3.5, the components of $\vec{F}_g$ are given by

$$\begin{aligned} \vec{F}_{g,x} &= |\vec{F}_g| sin\theta(\hat{\imath}), \\ \vec{F}_{g,y} &= |\vec{F}_g| cos\theta(-\hat{\jmath}), \end{aligned} \tag{3.1}$$

where the minus sign before the $\hat{\jmath}$ accounts for the fact that this component points in the negative y-direction.

[3] You may, of course, choose your axes to point exactly horizontally and vertically, in which case $\vec{F}_g$ would point entirely in the negative y-direction. In that case, the normal vector $\vec{F}_n$ would have both x- and y-components. But since other forces (such as friction and the delivery woman's push) generally point *along* the ramp surface, tilting your coordinate axes may well save you time later.

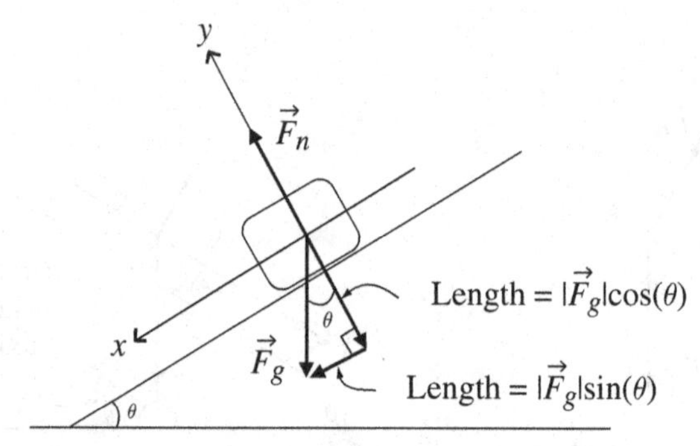

Figure 3.5 x- and y-components of $\vec{F}_g$.

A note about notation: as mentioned in Chapter 1, it's customary to write Eqs. 3.1 as

$$F_{g,x} = |\vec{F}_g| sin\theta,$$
$$F_{g,y} = -|\vec{F}_g| cos\theta, \tag{3.2}$$

that is, as scalars rather than vectors. That's because the direction of vector components should be clear from the subscript: the x-component is always in the $\hat{\imath}$ direction (or $-\hat{\imath}$ direction if it's negative), and the y-component is always in the $\hat{\jmath}$ direction (or $-\hat{\jmath}$ direction if it's negative). So you can write the components of a vector as scalars or vectors, as long as you remember that each component points in a specific direction, which means you cannot simply add the x- and y-components algebraically, even if they're written as scalars. You must add them as vectors.

Whether you write the components as vectors or scalars, having the x- and y-components of $\vec{F}_g$ in hand and knowing that the normal force of the plane on the box is entirely in the positive y-direction, you're now in a position to use vector addition to find the total force acting on the box. Writing the magnitude of the sum of the forces in the x-direction, you have

$$|\Sigma \vec{F}_x| = |\vec{F}_g| sin\theta, \tag{3.3}$$

and in the y-direction

$$|\Sigma \vec{F}_y| = (|\vec{F}_n| - |\vec{F}_g| cos\theta). \tag{3.4}$$

Alternatively, instead of writing separate equations for the x- and y-components of the total force, you can write a vector equation incorporating both:

$$\Sigma \vec{F} = (|\vec{F}_g| sin\theta)\hat{\imath} + (|\vec{F}_n| - |\vec{F}_g| cos\theta)\hat{\jmath}, \tag{3.5}$$

which contains exactly the same information as Eqs. 3.3 and 3.4.

Getting from the total force to the acceleration of the box is a simple step thanks to Isaac Newton, whose Second Law tells you that the magnitudes of the x- and y-components of the acceleration are

$$a_x = |\Sigma \vec{F}_x|/m = (|\vec{F}_g| sin\theta/m), \quad (3.6)$$

and

$$a_y = |\Sigma \vec{F}_y|/m = [(|\vec{F}_n| - |\vec{F}_g| cos\theta)/m], \quad (3.7)$$

or, in full vector form,

$$\vec{a} = \Sigma \vec{F}/m = (|\vec{F}_g| sin\theta/m)\hat{\imath} + [(|\vec{F}_n| - |\vec{F}_g| cos\theta)/m]\hat{\jmath}. \quad (3.8)$$

Whether you realize it or not, you almost certainly know two facts that will allow you to simplify these equations considerably. The first is that the magnitude of the force of gravity ($|\vec{F}_g|$) on an object of mass "m" is simply equal to mg, where "g" is the magnitude of the acceleration of gravity (9.8 m/s^2 at the Earth's surface).[4] So wherever you have the factor $|\vec{F}_g|$, you can substitute the expression mg.

The second simplification is produced by the realization that as long as the box stays on the ramp and doesn't fly off into the air or break through to the ground, the y-component of the acceleration (a_y) must remain zero (remember that the y-axis is perpendicular to the surface of the ramp). Using the fact that $|\vec{F}_g| = mg$ and that $a_y = 0$ turns Eqs. 3.6 and 3.7 into the following:

$$a_x = mg \sin\theta/m = g \sin\theta \quad (3.9)$$

and

$$a_y = (|\vec{F}_n| - mg \cos\theta)/m = 0. \quad (3.10)$$

When you're working a physics problem, it's a good idea to step back from your calculations once in a while to look at your intermediate results to see if they're trying to tell you something – and that's certainly the case at this point. Equation 3.9 already has an important result for you: in the absence of the upward-pushing delivery woman and with no friction, the box will accelerate down the ramp (that is, in the $+x$-direction) with an acceleration that depends on only two things: which planet the delivery truck is on (that is, the value of "g") and the angle that the ramp makes with the horizontal (θ). In this

[4] Remember that mass is a measure of the amount of material an object contains and weight is the force of gravity on that mass. So mass is a scalar (magnitude only) and weight is a vector (magnitude = mg and direction = straight down). Should you travel in space, your weight will change as you leave the Earth's gravity behind, but your mass will remain the same.

case, just as for a freely falling object, the mass of the box does not affect its acceleration.[5]

Since the sine of the ramp angle can never be greater than one, Eq. 3.9 also tells you that the magnitude of the acceleration of the object ($g \sin\theta$) can never be greater than g, the accleration of gravity. It can, of course, be equal to g if $\sin\theta = 1$. But this would mean that θ would have to be 90° (since $\sin 90° = 1$), in which case the ramp would be exactly vertical. In such cases, you no longer have an object sliding down a ramp, you have an object falling next to a wall.

There's also good information lurking in Eq. 3.10, but you have to think a bit to see it. According to this equation, the y-component of the box's acceleration is equal to the difference between the magnitude of the normal force ($|\vec{F}_n|$) and the y-component of the gravitational force ($mg \cos\theta$). But since you know that in this problem the box remains on the ramp and the y-acceleration is therefore zero, you can use Eq. 3.10 to determine the magnitude of the normal force. Since

$$a_y = (|\vec{F}_n| - mg \cos\theta)/m = 0,$$

then

$$|\vec{F}_n| = mg \cos\theta. \tag{3.11}$$

So the normal force depends on the weight of the object (mg) and the cosine of the ramp angle (θ). Understanding this will help you avoid a common pitfall for students who know that the normal force is the reaction force produced on the object by the ramp, and who then mistakenly conclude that the normal force must always equal the weight of the object (mg). That line of reasoning only works for *horizontal* surfaces, because for any inclined surface, it's only the component of the object's weight that's perpendicular to the surface that produces the reaction force we call the normal force. That perpendicular component of the object's weight is shown in Figure 3.5 to be $mg \cos\theta$, which spans the range from mg (when $\theta = 0°$, meaning the ramp is horizontal and bears the full weight of the object) to zero (when $\theta = 90°$, meaning the ramp is vertical and bears none of the object's weight). In all other cases, the magnitude of the normal force will have a value between 0 and mg.

If you're wondering why you should bother finding $\vec{F}_n$ if you're only interested in the x-component of the acceleration, the answer is that you may not

[5] But doesn't the Earth pull harder on a more-massive object? Yes it does, but a more-massive object also *resists acceleration* more than a less-massive object. Since gravitational mass (which determines how strongly gravity pulls on an object) has the same value as inertial mass (which determines how strongly the object resists acceleration), the result is that all objects fall freely (or slide freely down frictionless ramps) with an acceleration that does not depend on their mass.

care about $\vec{F}_n$ for the frictionless case (unless you're worried about your ramp breaking), but you'll definitely need $\vec{F}_n$ when friction exists between the ramp surface and the bottom of the box.

With the magnitude of the down-ramp component of the acceleration (a_x) available from Eq. 3.9, all that remains is for you to find the speed of the box at the bottom of the ramp. Finding speed from acceleration turns out to be quite straightforward, especially when the acceleration is constant (as it is in this case), provided that you're in possession of either one of two pieces of information: the time the box takes to reach the bottom of the ramp, or (more likely), the distance from the box's starting point to the bottom of the ramp. You'll also need the initial speed, which you can generally discern from the initial conditions, and which you can take to be zero in this case. As you may remember from kinematics, the final speed of an object moving in the x-direction with initial speed $v_{x,initial}$ undergoing constant accleration a_x over time t is given by

$$v_{x,final} = v_{x,initial} + a_x t, \tag{3.12}$$

or, if you know d, the distance in the positive x-direction over which the acceleration occurs,

$$(v_{x,final})^2 = (v_{x,initial})^2 + 2a_x d. \tag{3.13}$$

Using the expression for acceleration from Eq. 3.9, this becomes

$$(v_{x,final})^2 = (0)^2 + 2\,(g \sin\theta)\,d$$

or

$$v_{x,final} = \sqrt{2\,(g \sin\theta)\,d}. \tag{3.14}$$

So, for example, a box sliding down a 2 m ramp with an angle of 30° to the horizontal on the surface of the Earth will be moving at a speed of

$$v_{x,final} = \sqrt{2\left[(9.8\,\text{m/s}^2)\sin 30°\right]2\,\text{m}} = 4.4\,\text{m/s} \tag{3.15}$$

when it reaches the bottom of the ramp. If you're curious about how long it takes the box to travel the 2 m down the ramp under these conditions, you can plug this value for the final speed into Eq. 3.12 and solve for t, which turns out to be about 0.9 s in this case.

Stripping away effects such as friction is often a good way to learn the fundamentals of a problem, but if you've ever encountered a ramp outside of physics texts, there's a good chance you had to deal with friction. Happily, once you understand how to use vectors, including friction in the "box on a ramp" problem becomes a simple matter of adding another force into the mix before solving for the acceleration.

As you may recall, friction operates in two regimes: "static" friction determines how hard you have to push on a stationary object to get it moving, but once the object is moving, the frictional force that opposes the motion is produced by "kinetic" friction. So although both types of friction oppose motion, the magnitude of the force produced by static friction depends on the applied force (the harder you push, the stronger the opposing force of static friction, until the object "breaks free" and begins moving), while the magnitude of the kinetic-friction force depends only on the normal force and the coefficient of kinetic friction between the object and the surface.[6] To determine the effect of kinetic friction on the speed of the box at the bottom of the ramp, you can modify your free-body diagram to include the frictional force ($\vec{F}_f$), as shown in Figure 3.6.

Notice that the direction of the frictional force is chosen so as to oppose the motion, and since the box is moving down the ramp in this case, the force of kinetic friction points up the ramp (in the negative x-direction).

To determine the effect of friction on the acceleration of the box sliding down the ramp, you simply have to include the frictional force ($\vec{F}_f$) in your equation for the sum of the forces in the x-direction (Eq. 3.3), which becomes

$$|\Sigma \vec{F}_x| = |\vec{F}_g| sin\theta - |\vec{F}_f|. \tag{3.16}$$

This makes the acceleration

$$a_x = \Sigma F_x/m = \left(|\vec{F}_g| sin\theta - |\vec{F}_f|\right)/m. \tag{3.17}$$

Clearly, to determine the magnitude of the acceleration (a_x), you'll need to find an expression for $|\vec{F}_f|$, just as you used $mg \sin\theta$ for $|F_{g,x}|$ in Eq. 3.9.

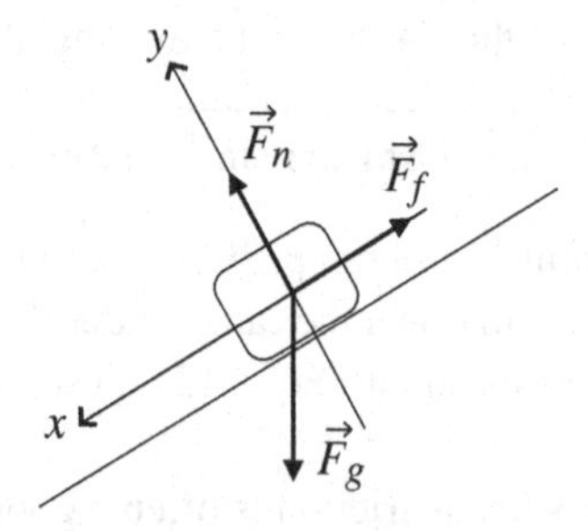

Figure 3.6 Free-body diagram for object on ramp with friction.

[6] You can read more about this in introductory physics texts such as Serway & Jewett or Halliday, Resnick, & Walker.

Fortunately, that's easy to do, because the magnitude of the force of kinetic friction is simply the product of the magnitude of the normal force ($|\vec{F}_n|$) and the coefficient of kinetic friction (μ_k):

$$|\vec{F}_f| = \mu_k |\vec{F}_n|. \tag{3.18}$$

You also know from Eq. 3.11 that $|\vec{F}_n| = mg \cos\theta$, so

$$\begin{aligned} a_x &= (mg \sin\theta - \mu_k mg \cos\theta)/m \\ &= (g \sin\theta - \mu_k g \cos\theta). \end{aligned} \tag{3.19}$$

Comparing this expression for the acceleration of the box to the acceleration in the frictionless case (Eq. 3.9), you'll be happy to note that the term due to gravity ($g \sin\theta$) is exactly the same in both cases, and the term due to friction ($\mu_k g \cos\theta$) is *subtracted* from the gravity term. This means that the acceleration of the box will be made smaller by the frictional force. So in the case considered previously of a box sliding down a 2 m ramp that makes an angle of 30° with the horizontal, if the coefficient of kinetic friction between the box and the ramp is 0.4, the speed of the box at the bottom of the ramp will be reduced to

$$\begin{aligned} v_{x,final} &= \sqrt{2\left[(9.8\,\mathrm{m/s^2})\sin 30^\circ - (0.4)(9.8\,\mathrm{m/s^2})\cos 30^\circ\right]2\,\mathrm{m}} \\ &= 2.5\,\mathrm{m/s}. \end{aligned} \tag{3.20}$$

There is one aspect of Eq. 3.19 that may worry you: what if the second term is larger than the first? For any angle between 0° and 45°, the cosine is bigger than the sine, so if the coefficient of kinetic friction (μ_k) is sufficiently large, this equation predicts that the acceleration will be in the *negative* x-direction, meaning the box will acclerate up the ramp even if no one is pushing on it. As physicists like to say, "That's not physical," meaning that this result contradicts other well-established laws of physics (conservation of energy comes to mind in this case). So where have we gone wrong in our analysis? We haven't, really, you just need to think carefully about the initial assumptions. One of those assumptions was that the box is travelling *down* the ramp, which is why we drew the frictional force pointing up the ramp in our free-body diagram (Figure 3.6). But if the ramp isn't very steep and the coefficient of friction between the box and the ramp is sufficiently large, the down-ramp component of the force of gravity will not be strong enough to overcome the frictional force, and the box will not slide down the ramp.[7] So there's nothing wrong

[7] You can determine whether the box will move by comparing the maximum static frictional force (which is just the product of the coefficient of static friction and the normal force) to the sum of the x-components of all the other forces.

with Eq. 3.19, it's just that it only applies to the situation in which the box is moving down the ramp under the influence of gravity, in which case the force of kinetic friction points up the ramp.

So there you have it. You've used vectors to represent the forces of gravity and friction, and knowing how to find vector components and how to perform vector addition has allowed you to find the acceleration and speed of the box under various conditions. And if a box sliding straight down a ramp is a bit too mundane for your taste, you may want to take a look at the next three application examples. In them, you'll see how vectors can be helpful in analyzing motion on a curved path and how vector operations can be used to understand the behavior of electric and magnetic fields.

3.2 Curvilinear motion

In everyday language, the word "acceleration" is used as a synonym for "increasing speed." Hence the "accelerator" in an automobile usually refers to the gas pedal. But in physics and engineering, acceleration is defined as any change in velocity, and velocity is a vector quantity with both magnitude and direction. So changing the *direction* of the velocity is also a form of acceleration, meaning that most cars have three accelerators: the gas pedal, the brake, and the steering wheel. "Stepping on the gas" produces an acceleration in the same direction as the velocity vector (causing the speed to increase), pressing on the brake produces an acceleration directly opposite to the direction of the velocity vector (causing the speed to decrease), and turning the steering wheel produces an acceleration perpendicular to the velocity vector (causing the car's direction to change but not affecting the speed).[8] Acceleration in the direction parallel (or antiparallel) to the velocity vector is called "tangential" and acceleration perpendicular to the velocity is called "radial." Any time an object experiences radial acceleration, it does not move in a straight line, and its motion is called "curvilinear." An example of curvilinear motion is shown in Figure 3.7, in which a car is going around a curve.

Note that at any instant, the velocity vector points directly along the path the car is following. For a curving path, that means the instantaneous velocity vector is tangent to the path, as you can see when the car is at position B in Figure 3.7. If you wish to determine the acceleration at points such as A, B, and C along the car's path, it's not enough to know the velocity at those points; you have to know how the velocity is *changing* with time at those locations.

[8] In reality, turning the steering wheel produces frictional forces that also slow the car down, but it's the perpendicular component of the acceleration that causes the car to turn.

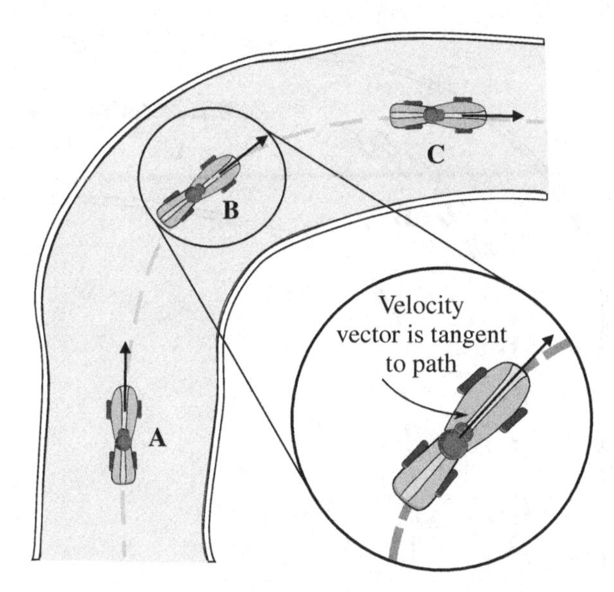

Figure 3.7 Velocity vectors for a car following a curved path.

A good way to visualize the acceleration vector is to graphically represent the velocity vector at the instants of time just before and just after the car is at positions A, B, and C. This is illustrated in Figure 3.8 for the following case: the car is slowing down at Position A as it approaches the turn, maintaining constant speed while turning at Position B, and then speeding up as it exits the turn at Position C.

You can get a sense of the acceleration just by examining the change in the velocity vectors at each position. Comparing the velocity vectors just before and just after Position A, you can see that the magnitude (length) of the vector is getting smaller but the direction remains the same. This means that the speed of the car is decreasing but the car is not yet turning. Now look at the velocity vectors just before and after Position B: the direction of the vector is changing but its length is not, so the car is turning while maintaining constant speed. Finally, by examining the velocity vectors before and after Position C, you can see that the length is increasing, meaning the car is speeding up after leaving the turn.

The direction of the acceleration is easily found by remembering that the average acceleration is given by the equation $\vec{a} = \Delta\vec{v}/\Delta t$, where $\Delta\vec{v}$ is the change in velocity over time Δt. That change in velocity is just $\vec{v}_{final} - \vec{v}_{initial}$, which you can determine by subtracting the earlier velocity vector from the later one at each position in Figure 3.8. To make that easier, the vectors are reproduced in Figure 3.9.

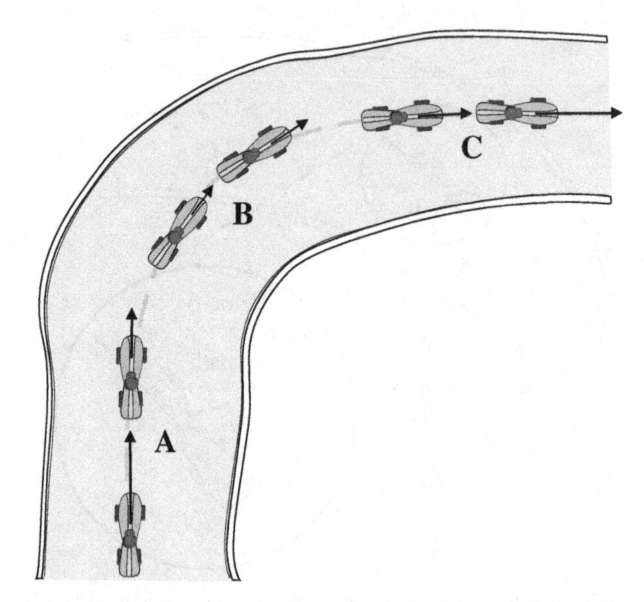

Figure 3.8 Change in car's velocity vectors at Positions A, B, and C.

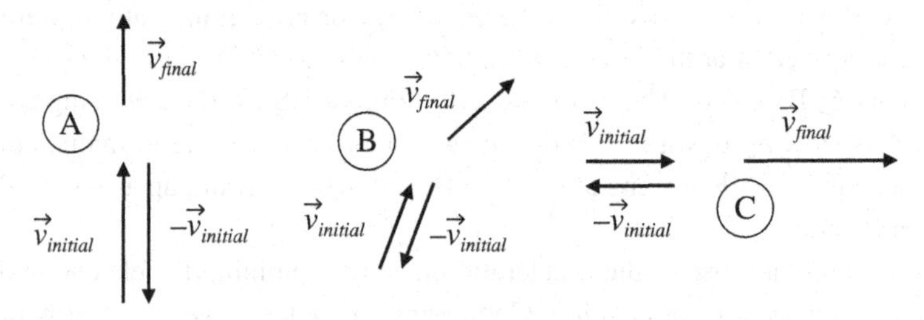

Figure 3.9 Velocity vectors before and after Positions A, B, and C.

Note that the vectors shown in Figure 3.9 include not only $\vec{v}_{final}$ and $\vec{v}_{initial}$, but also the negative of $\vec{v}_{initial}$. That's because you'll need to know $-\vec{v}_{initial}$ to compute the change in velocity, since $\Delta\vec{v} = \vec{v}_{final} - \vec{v}_{initial}$, which is the same as $\vec{v}_{final} + (-\vec{v}_{initial})$. Remember that to add two vectors graphically you simply move the tail of one to the head of the other and then draw the resultant from the start of the first to the end of the second vector. The results of adding vectors $\vec{v}_{final}$ and $-\vec{v}_{initial}$ are shown in Figure 3.10.

In Figure 3.10, the velocity vectors $-\vec{v}_{initial}$ and $\vec{v}_{final}$ for Positions A and C are shown slightly offset since they would overlay one another if they were drawn truly head-to-tail. If you look at the direction of the vector representing the change in velocity ($\Delta\vec{v}$) at each position, you'll see that while the car is

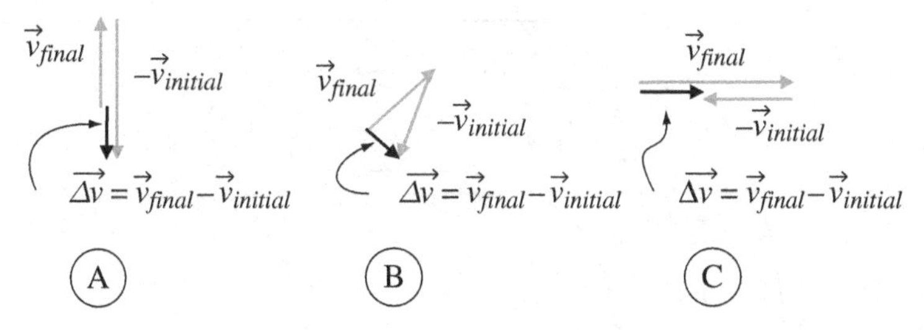

Figure 3.10 Change in velocity vectors at Positions A, B, and C.

slowing down at Position A, the change in velocity is in the opposite direction from the velocity at this point. Since the acceleration ($\vec{a}$) is defined as the vector change in velocity ($\Delta\vec{v}$) divided by the scalar time period (Δt) over which that change occurs, the direction of $\vec{a}$ must be the same as the direction of $\Delta\vec{v}$. Hence the acceleration direction at Position A is opposite to the direction of the velocity vector, as you'd expect when the car is slowing down. This is an example of negative tangential acceleration.

Now consider the direction of the vector change in velocity $\Delta\vec{v}$ at Position B, where the car is going around the turn at constant speed. In this case, subtracting $\vec{v}_{initial}$ from $\vec{v}_{final}$ gives a vector $\Delta\vec{v}$ that is *perpendicular* to the velocity vector. This shows that the acceleration vector for an object moving along a curve at constant speed points toward the center of curvature (to help you visualize this direction, the $\Delta\vec{v}$ vectors are shown on the car's path in Figure 3.11). At position B, this is an example of radial acceleration.[9]

Finally, as the car speeds up at Position C, you can see that the direction of the vector change in velocity $\Delta\vec{v}$ is the same as the direction of the velocity vector, meaning that the accleration in this case is parallel to the velocity. Hence this is an example of positive tangential acceleration.

For Position B, a careful analysis of the length of the vector change in velocity reveals that the magnitude of the radial acceleration depends on the square of the speed and on the radius of curvature of the path. Before getting into that, it's worth a few minutes of your time to make sure you understand the terminology commonly used to describe acceleration and force in curvilinear motion. Acceleration toward the center of curvature (such as the acceleration at Position B in Figure 3.11) is called "centripetal" (for "center-seeking") acceleration, and the force producing that acceleration is often called centripetal force. It's important for you to understand that a centripetal force is not a new

[9] As described later in this section, most texts define the positive direction for radial acceleration to be outward from the center of curvature, in which case the acceleration at Point B would be considered negative radial acceleration.

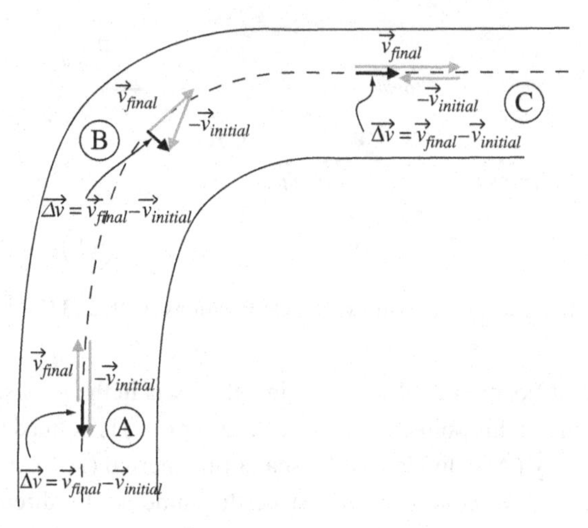

Figure 3.11 Acceleration vectors at Positions A, B, and C.

kind of force that is somehow different from mechanical, electrical, magnetic, or other kinds of force. The word "centripetal" simply describes the *direction* of the force, but the force itself is provided by the same old kinds of forces to which you're accustomed. So for a car going around a curve, the centripetal force is simply the frictional force of the tires on the ground. If you tie a rock to a rope and twirl the rope in a circle, the centripetal force on the rock is produced by the tension of the rope. And if you fill a bucket with water and swing it over your head, the centripetal force on the bucket (and via the bucket on the water) comes from the muscles in your arm. So the centripetal force is whatever force is producing the centripetal acceleration that causes the object to follow a curved path.

As footnoted earlier, it's conventional to consider radial acceleration ($\vec{a}_r$) as positive outward (*away* from the center of curvature), and since centripetal acceleration ($\vec{a}_c$) is defined as positive *toward* the center of curvature, you may run across an equation such as $\vec{a}_r = -\vec{a}_c$. This is simply a statement that the radial acceleration and centripetal acceleration are commonly defined to have the same magnitude but opposite directions.

You should note that in the case of the car on the curving road, the rock being twirled in a circle on a rope, and the bucket of water being swung over your head, the centripetal acceleration (and hence the centripetal force) is toward the center of curvature, and there is no acceleration (and no force) pointing radially outward. But what about the "centrifugal" force that the occupants of the car feel toward the *outside* of the curve (that is, toward the left door if the car is turning to the right)? What they're feeling is the force of the left door on their bodies as they attempt to obey Newton's First Law and continue moving in a straight line while the car is accelerating to the right. So centrifugal force is the apparent force experienced by observers in the reference frame that is rotating with the object (physicists refer to acclerating reference frames such as this as "non-inertial"). Hence if you're riding in a right-turning car, as you slide across the seat and up against the left door, in your (rotating) reference frame you're accelerating to your left, which causes you to conclude that there's a force in that direction (outward from the center of curvature). But for those of us not riding in the car, we don't see any such force; we simply observe the centripetal acceleration of the car as the friction of the tires on the road provides a centripetal (rightward) force.

The concept of centripetal and centrifugal force can be understood by considering an Olympic hammer-thrower as she spins a heavy mass on the end of cable, as illustrated from above in Figure 3.12. For the thrower, it feels like the object is pulling directly outward (away from her). Once again, in the non-rotating reference frame of the stadium, that's just because the object is attempting to obey Newton's First Law and continue moving in a straight line. So from our vantage point in the viewing stand, we see that the hammer-thrower is having to produce a centripetal (radially inward) force to make the object follow a curved path.

So is the hammer-thrower wrong in her assessment? Absolutely not. In her reference frame, which is rotating along with the mass, her conclusion that a radially outward (centrifugal) force exists is perfectly valid. After all, she knows that she has to exert a very strong inward force on the cable to keep the mass at the same distance from her (because in her reference frame the mass

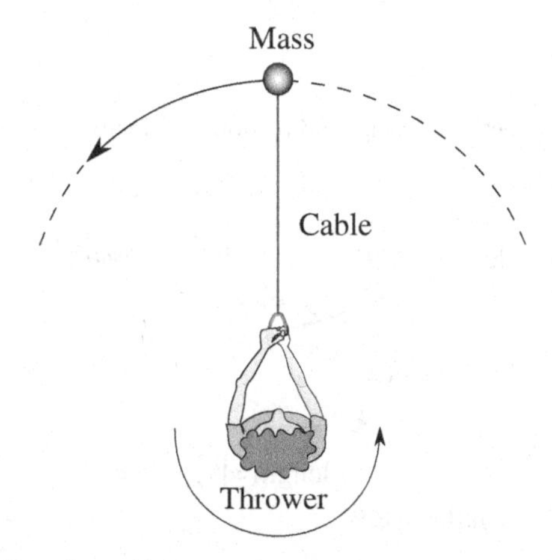

Figure 3.12 Top view of hammer-thrower.

has zero acceleration until she releases it). Hence she is correct in concluding that in her reference frame there must be a force in the radially outward direction to balance her inward pull. So if you hear someone say that the centrifugal force is "fictitious," they generally mean that centrifugal force is an apparent force to an observer in a rotating (non-inertial) reference frame.

Once you understand the concepts of centripetal acceleration and force, it's reasonable to ask how strong the centripetal force must be to cause an object to follow a given path. It's simple to determine the centripetal force using Newton's Second Law ($\vec{F} = m\vec{a}$) if you know the object's mass and have some way of finding the centripetal acceleration. Happily, the centripetal acceleration turns out to depend only on the object's speed and the radius of curvature of the path, as you can see by considering Figures 3.13 and 3.14.

In Figure 3.13 you can see the velocity vectors at two locations for an object in uniform circular motion (meaning that the object's speed and the radius of curvature are both constant over the time period under consideration). Note that the two positions are separated by angle $\Delta\theta$ at the center of curvature, which makes the arc length between the initial and final positions equal to $r\Delta\theta$, where r is the radius of curvature and $\Delta\theta$ is in radians. Since the speed of the object is constant over this distance, you know that $|\vec{v}_{initial}|$ must equal $|\vec{v}_{final}|$ (in other words, the direction but not the length of the velocity vector

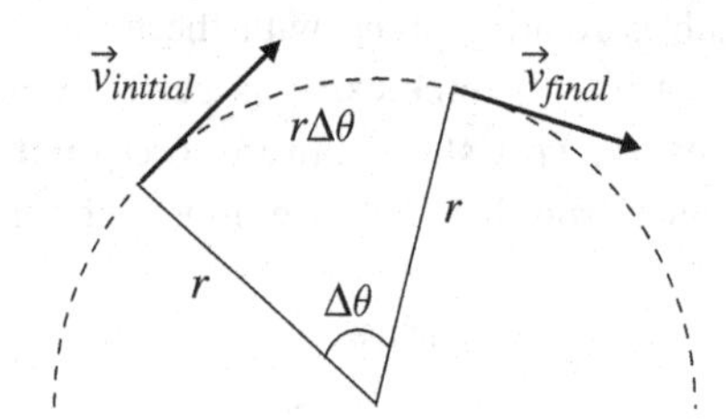

Figure 3.13 Geometry of changing direction of velocity.

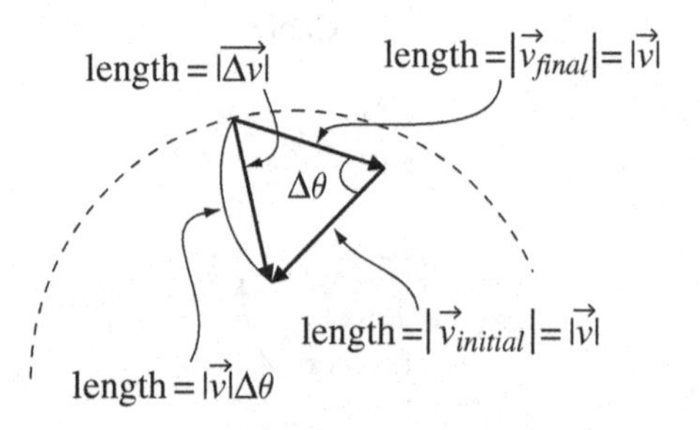

Figure 3.14 Geometry for determining length of $\Delta\vec{v}$.

has changed). You can therefore set $|\vec{v}_{initial}| = |\vec{v}_{final}| = |\vec{v}|$, where $|\vec{v}|$ is the speed of the object at both positions. Since the average speed of the object is defined as the distance covered divided by the time taken to cover that distance, you can write

$$|\vec{v}| = \frac{r\Delta\theta}{\Delta t}, \tag{3.21}$$

which means that

$$\Delta\theta = \frac{|\vec{v}|\Delta t}{r}. \tag{3.22}$$

The reason that an expression such as Eq. 3.22 for $\Delta\theta$ is valuable is that this angle change is directly related to the magnitude of the vector change in velocity, which you need to know if you want to find the centripetal acceleration. To see that, consider what happens if you form the vector $\Delta\vec{v}$ by adding $\vec{v}_{final}$ to $-\vec{v}_{initial}$, as in Figure 3.14. The first thing you should note is that the angle between the vectors $\vec{v}_{final}$ and $-\vec{v}_{initial}$ is equal to $\Delta\theta$ (if you don't see why that's true, go back to Figure 3.13 and imagine extending both vectors $\vec{v}_{final}$ and $-\vec{v}_{initial}$ until they cross). Also note that the vector $\Delta\vec{v}$ is drawn at the location mid-way between the original location of $\vec{v}_{initial}$ and the original location of $\vec{v}_{final}$, since that's the location at which you're finding the centripetal acceleration. The final thing to note in this figure is that both $\vec{v}_{final}$ and $-\vec{v}_{initial}$ have length equal to $|\vec{v}|$, which makes the arc length shown in the figure equal to $|\vec{v}|\Delta\theta$.

Now imagine what will happen if you allow the angle $\Delta\theta$ to shrink toward zero. As the angle decreases, the arc length $|\vec{v}|\Delta\theta$ will get closer and closer to the length of $\Delta\vec{v}$. Plugging in the value for $\Delta\theta$ from Eq. 3.22, you have in the small-angle limit

$$\begin{aligned} |\Delta\vec{v}| &\approx |\vec{v}|\Delta\theta = |\vec{v}|\frac{|\vec{v}|\Delta t}{r} \\ &= \frac{|\vec{v}|^2\Delta t}{r}, \end{aligned} \tag{3.23}$$

which means that the magnitude of the instantaneous centripetal acceleration is

$$\begin{aligned} |\vec{a}_c| &= \frac{|\Delta\vec{v}|}{\Delta t} = \frac{|\vec{v}|^2\Delta t}{r\Delta t} \\ &= \frac{|\vec{v}|^2}{r}. \end{aligned} \tag{3.24}$$

So there you have it: the centripetal acceleration at any given point is simply the square of the speed divided by the radius of curvature of the path at that point. Hence doubling your speed means that your centripetal acceleration is *four* times larger, which means that the centripetal force must be four times stronger.

If you're concerned that Eq. 3.24 may apply only in the case of uniform circular motion, remember that by allowing $\Delta\theta$ to become arbitrarily small you've ensured that neither the speed nor the radius of curvature has changed during the time period under consideration.

What does Eq. 3.24 tell you about the amount of force needed to cause an object to follow a specified curving path? Consider the hammer-thrower discussed above and shown in Figure 3.12, and assume that she intends to launch a 4 kg mass at the end of a 1.2 m cable with a speed of 20 m/s. Assuming she achieves her maximum speed just before letting go of the cable, at that point the centripetal acceleration will be

$$|\vec{a}_c| = \frac{|\vec{v}|^2}{r} = \frac{(20\,\text{m/s})^2}{1.2\,\text{m}}$$
$$= 333.3\,\text{m/s}^2,$$

which means that the thrower must provide a centripetal force of

$$|\vec{F}_c| = m|\vec{a}_c| = 4\,\text{kg}\left(333.3\,\text{m/s}^2\right)$$
$$= 1333.3\,\text{N}$$

which is almost 300 pounds of force (and this doesn't include the mass of the cable).

With Eq. 3.24 to help you find the magnitude of the centripetal acceleration, and knowing that the tangential acceleration is just the change in speed over time ($\vec{a}_{tang} = \Delta\vec{v}/\Delta t$), the total acceleration can be found through vector addition, as shown in Figure 3.15. Thus the magnitude of the total acceleration is

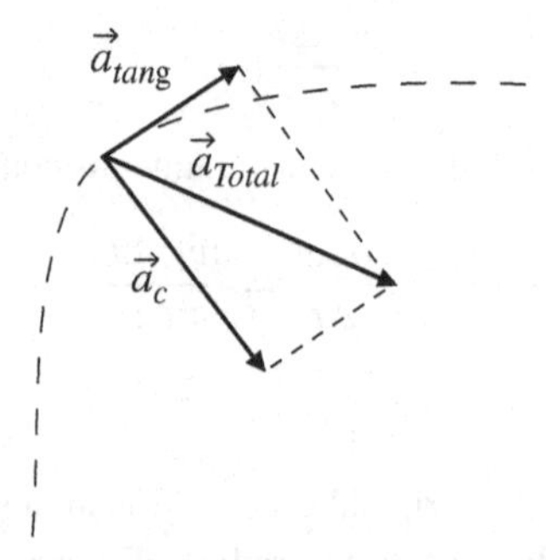

Figure 3.15 Total acceleration as the vector sum of centripetal and tangential acceleration.

$$|\vec{a}_{Total}| = \sqrt{(|\vec{a}_c|)^2 + (|\vec{a}_{tang}|)^2}$$
$$= \sqrt{\left(\frac{v^2}{r}\right)^2 + \left(\frac{|\Delta\vec{v}|}{\Delta t}\right)^2}. \quad (3.25)$$

You'll find an example of combined tangential and centripetal acceleration in the problems at the end of this chapter.

3.3 The electric field

If the previous two sections convinced you that vectors are very helpful in solving mechanics problems, the next two sections should help you understand why vectors are absolutely essential in problems involving electric and magnetic fields and their effect on charged particles. You'll also see how the vector operations of divergence, curl, gradient, and Laplacian are used in electrostatics. Even if you've never taken an E&M course (and never hope to), the examples in these sections should be sufficiently self-contained to allow you to understand how vectors and vector operations can be used in E&M.

The natural way to begin a discussion of electric and magnetic fields is to provide a clear, concise definition that states exactly what an electric or magnetic field *is*. Such a definition would appear right here if I had one. But almost two centuries after Michael Faraday first used the words "field of force" to describe the region around electric charges, we still don't have a standard way of saying what such a field is. The Oxford English Dictionary provides definitions for "field" that include an "area or space" under the influence of an agent, a "state or situation" in which force is exerted, and the "action" of a force. According to James Clerk Maxwell, "The electric field is the portion of space in the neighbourhood of electrified bodies." In Halliday, Resnick, and Walker you can learn to define the electric field by placing a small positive test charge q_0 at some point and measuring the electrostatic force $\vec{F}_E$ on that test charge;[10] the electric field $\vec{E}$ is then defined as $\vec{E} = \vec{F}_E/q_0$. In Griffiths' *Introduction to Electrodynamics*, he states that "... physically, $\vec{E}(P)$ is the force per unit charge that would be exerted on a test charge placed at P." The words "would be" in that definition are important, because it is *not* necessary for the test charge to be present in order for the field to exist.

[10] Why do physics and engineering texts always refer to a *small* test charge? For two reasons: firstly, the amount of charge on the test charge must be small so that the electric field produced by the test charge is negligible when compared to the electric field that you're trying to determine using the test charge. Secondly, the test charge must be physically small because you're using it to determine the field at a specific position, so you don't want your test charge to extend over a large region of space.

The common thread running through all these definitions is this: fields and forces are closely related. So we'll take the following as our definition of the electric field $\vec{E}$:

$$\vec{E} \equiv \frac{\vec{F}_E}{q_0}, \tag{3.26}$$

where $\vec{E}$ is the vector electric field, q_0 is a small test charge, and $\vec{F}_E$ is the electric force produced on the test charge by the electric field. Defining the electric field through this equation should help you remember that $\vec{E}$ is a vector quantity with magnitude directly proportional to force and with direction given by the direction of the force on a positive test charge (because if q_0 is negative, there would be a minus sign on one side of the equation, which would mean that vector $\vec{E}$ would be in the opposite direction from vector $\vec{F}_E$).

This definition should also help you see that $\vec{E}$ has dimensions of force divided by charge, for which the standard (SI) units are newtons per coulomb (N/C). These units are equivalent to volts per meter (V/m), since volts have dimensions of force times distance divided by charge (units of newtons times meters/coulombs). So you'll find the units of electric field given as N/C in some texts and V/m in others, and you can rest assured that these mean exactly the same thing.

There is, however, something important to be noticed in the units of the electric field vector: the dimension of length (units of meters in this case) appears in the *denominator* of the dimensions of the electric field. And that means that the vector that represents an electric field has a fundamental difference from the vectors that represent quantities such as position (which has dimension of length), velocity (dimension of length over time), or acceleration (dimension of length over time squared). As you can read in Chapter 4, that's because vectors whose dimensions contain length in the numerator transform oppositely to vectors whose dimensions have length in the denominator when you perform certain coordinate-system changes. If this seems unclear and you don't plan to venture into the tensor portion of this book, do not panic; none of this will prevent you from using the concepts and operations described in Chapters 1 and 2 to solve problems involving vectors of this kind, exactly as you're about to do in the remainder of this section. But if you've run across objects called "one-forms" or "covectors" (of which the electric field is an example) and you're wondering how those objects are different from the things you've been calling vectors, the appearance of length in the denominator of the dimension is the beginning of the answer (you'll find the rest of the answer in Chapter 4 if you're interested).

You should also make sure you understand that if you know the electric field $\vec{E}$ at a given location, placing any amount of charge q at that location will result in an electric force $\vec{F}_E$ given by

$$\vec{F}_E = q\vec{E}. \tag{3.27}$$

So while Eq. 3.26 uses the electric force on a positive test charge to *define* the electric field, Eq. 3.27 is a generally useful expression for finding the electric force on any amount of charge at the location for which the electric field is known.

Defining an electric field is useful, but exactly how would you go about *producing* an electric field? One way is to gather up some electric charge, because every bit of charge produces an electric field, just as every bit of mass produces a gravitational field. Electric fields can also be produced by changing magnetic fields, but it is the "electrostatic" field produced by stationary electric charge that will be used to demonstrate the application of vectors in this section.

It's often helpful to be able to visualize the electric field in the vicinity of a charged object. The most common approaches to constructing a visual representation of an electric field are to use either arrows or "field lines" which point in the direction of the field at each point in space. In the arrow approach, the strength of the field is indicated by the length of the arrow, while in the field-line approach, it's the density of the lines that tells you the field strength, with closer lines signifying a stronger field. When you look at a drawing of electric field lines or arrows, be sure to remember that the field exists between the lines as well.

The electric fields produced by positive and negative point charges are shown using the arrow approach in Figure 3.16 and using the field-line approach in Figure 3.17. When you look at electric field lines such as these,

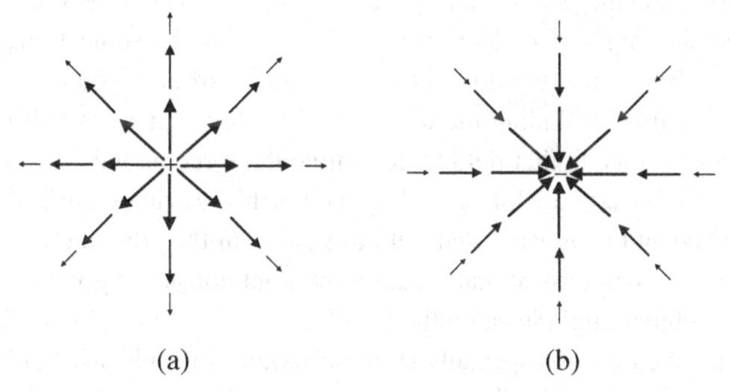

Figure 3.16 The electric field of positive and negative point charges drawn using arrows.

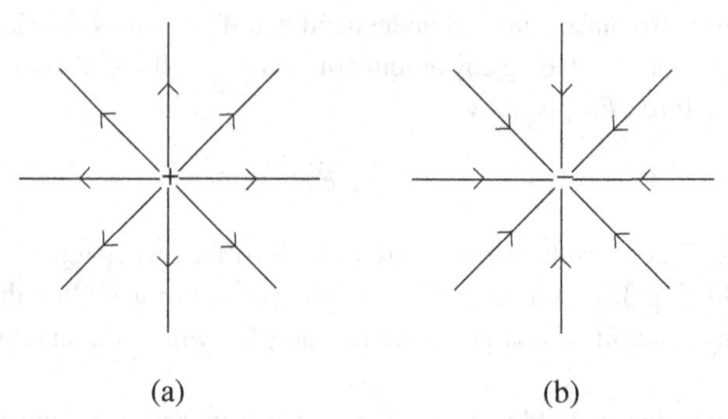

Figure 3.17 The electric field of positive and negative point charges drawn using field lines.

don't forget that the field arrows and lines always point in the direction of the electric force on a *positive* test charge, and that electrostatic field lines always begin on positive charge and end on negative charge. And since the field lines show the direction of the electric field at any given point, it's impossible for two fields lines to cross, since that would indicate that the electric field is pointing in more than one direction at the point of intersection (if two electric fields are superimposed at a given point, they simply add as vectors to give the total electric field at that point, and that total field can only point in a single direction).

At this point, you should make sure that you understand that electric fields can both *be produced* by electric charge as well as *produce* a force on another electric charge. So you're likely to face problems in which you first have to determine the total electric field produced by charge at a certain location and then figure out the effect of that field on a completely different charge (not one of the charges producing the field). But doesn't the charge that's being affected (let's call that one the "subject charge") also produce its own electric field? Yes it does, but as long as the electric field produced by the subject charge isn't strong enough to cause the other charges to move around, you can approach problems like this by finding the total electric field produced by all the other charges and then using that field to determine the force on the subject charge. This approach is very much like finding the Earth's gravitational field at some point in space and then using that field to figure out the gravitational force on an object of known mass at that location, without considering what effect the mass of the object might have on the Earth.

Problems like this are especially straightforward if the electric field is being produced by one or more discrete point charges. That's because the electric field $\vec{E}$ of a point charge q is simply

$$\vec{E} = k_e \frac{q}{r^2} \hat{r}, \tag{3.28}$$

where k_e is the Coulomb constant ($8.99 \times 10^9\,\mathrm{Nm^2/C^2}$), r is the distance in meters from the point charge to the location at which the electric field is being determined, and $\hat{r}$ is a unit vector pointing radially outward from the point charge.

Thus a single proton (electric charge of 1.6×10^{-19} C) at a distance of one meter produces an electric field given by

$$\begin{aligned}\vec{E} &= (8.99 \times 10^9\,\mathrm{Nm^2/C^2})\left(\frac{1.6 \times 10^{-19} C}{(1\,\mathrm{m})^2}\right)\hat{r}\\ &= 1.45 \times 10^{-9} (\mathrm{N/C})\,\hat{r}.\end{aligned}$$

Note that the direction of that field is radially *away* from the proton, since the unit vector $\hat{r}$ always points radially outward from the origin. An electron, having negative charge, produces an electric field of the same magnitude as that of the proton, but the electron's electric field points *toward* the electron. To see that, note that when you plug in a negative charge for q in Eq. 3.28, you have

$$\begin{aligned}\vec{E} &= (8.99 \times 10^9\,\mathrm{Nm^2/C^2})\left(\frac{-1.6 \times 10^{-19}\,\mathrm{C}}{(1\,\mathrm{m})^2}\right)\hat{r}\\ &= -1.45 \times 10^{-9} (\mathrm{N/C})\,\hat{r} = 1.45 \times 10^{-9} (\mathrm{N/C})\,(-\hat{r}),\end{aligned}$$

where the minus sign tells you that the direction of the electron's electric field is in the *negative* $\hat{r}$ direction, which is *toward* the source charge (since $\hat{r}$ is always radially outward, *minus* $\hat{r}$ is always radially inward). This is consistent with electric field lines beginning on positive charge and ending on negative charge.

To understand how to add the vector electric fields, consider the situation shown in Figure 3.18. Note that q_1 is positive, so its electric field must point radially outward from the location of q_1, while q_2 and q_3 are negative, so their

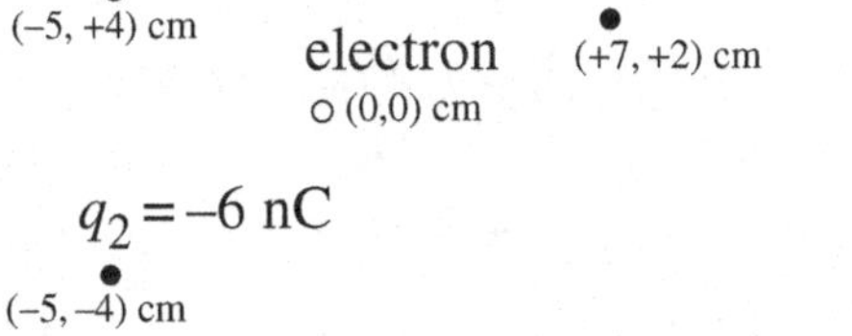

q_1 = +5 nC
(–5, +4) cm
q_3 = –8 nC
(+7, +2) cm
electron
o (0,0) cm
q_2 = –6 nC
(–5, –4) cm

Figure 3.18 Example values for charges near an electron.

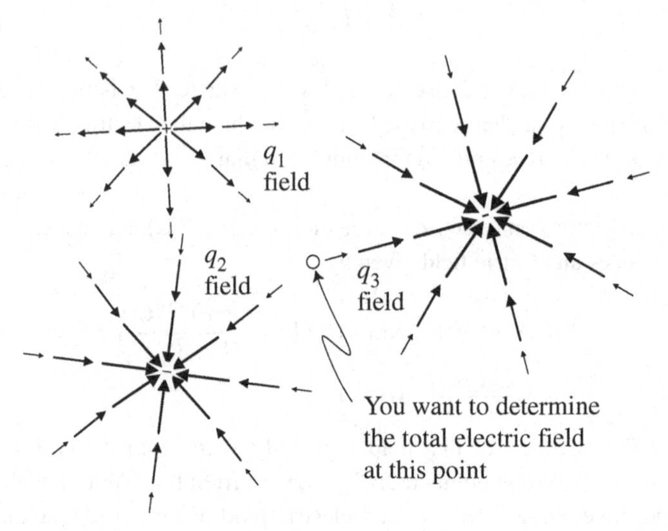

Figure 3.19 The electric fields produced by charges q_1, q_2, and q_3.

electric fields must point radially inward toward their locations. To find the total electric field at the position of the electron, it may help you to picture the fields produced by q_1, q_2, and q_3 as shown in Figure 3.19.

If you read the discussion of field lines earlier in this section, you should realize that the electric field exists *between* the lines as well as at the locations of the lines themselves. But just to help you visualize the direction of the fields from each of the three charges, the field lines in Figure 3.19 have been drawn on a tilt so that they are directly in line with the location at which you're trying to find the total field (the origin in this case). You should also remember that just because the lines have grown too small to see does not mean that the field has gone to zero. Hence the electric field produced by q_1 points down and to the right at the location of the electron, the field from q_2 points down and to the left, and the field from q_3 points up and to the right. It is these three vector fields that you will have to add together to determine the total electric field at the point of interest.

Using Eq. 3.28, the electric fields due to the three point charges q_1, q_2, and q_3 may be written as

$$\vec{E}_1 = k_e \frac{q_1}{r_1^2} \hat{r}_1,$$
$$\vec{E}_2 = k_e \frac{q_2}{r_2^2} \hat{r}_2, \tag{3.29}$$
$$\vec{E}_3 = k_e \frac{q_3}{r_3^2} \hat{r}_3.$$

Of course, you know from Figure 3.19 that these three electric fields do not point in the same direction. That's because the unit vector $\hat{r}_1$ points radially outward from the location of charge q_1, and $\hat{r}_2$ and $\hat{r}_3$ point radially outward from q_2 and q_3, respectively. This means you can't add the three electric fields algebraically; to find the total field you must use vector addition. You'll find an example of the vector addition of electric fields in the problems at the end of this chapter and the on-line solutions.

As you might suspect, it's not just the simple operations of vector addition and multiplication by a scalar that find use in electrostatics. If you followed the discussion of the divergence operation in Chapter 2, you may be wondering about the divergence of the electrostatic fields produced by a point charge (Figures 3.16 and 3.17). In fact, one of the fundamental laws of electrostatics is Gauss's Law for electric fields, the differential form of which is

$$\vec{\nabla} \circ \vec{E} = \rho/\epsilon_0, \tag{3.30}$$

where ρ represents the volume electric charge density (coulombs per cubic meter) and ϵ_0 is the vacuum permittivity of free space (8.85×10^{-12} Nm^2/C^2).

Gauss's Law for electric fields tells you that electric field lines diverge from any location at which positive charge exists (positive ρ) and converge upon any location at which negative charge is present (negative ρ). This explains the analogy between the "flow" of electrostatic field lines and the flow of a fluid. In this analogy, positive charge acts as the "source" of electrostatic field lines in the same sense as a faucet acts as the source of fluid, and negative charge acts as a "sink" of electrostatic field lines just as a drain does for fluid.

Note what happens when you take the divergence of the electric field of a point charge (this is most easily done in spherical coordinates):

$$\begin{aligned}\vec{\nabla} \circ \vec{E} &= \frac{1}{r^2}\frac{\partial}{\partial r}(r^2 E_r) = \frac{1}{r^2}\frac{\partial}{\partial r}\left(r^2 k_e \frac{q}{r^2}\right)\\ &= \frac{1}{r^2}\frac{\partial}{\partial r}(k_e q) = 0.\end{aligned}$$

This is consistent with the worked example in Chapter 2 showing that the divergence of any radial vector field is zero if the amplitude of the field falls off as $1/r^2$. Zero, that is, at all locations except where $r = 0$, the location of the source of the field. Thus Gauss's Law tells you that electrostatic field lines diverge only from those locations at which positive electric charge exists, and converge only on those locations at which negative charge exists.

You can gain additional understanding of the behavior of the electrostatic field by considering the curl of $\vec{E}$ for a point charge. Since E_θ and E_ϕ are both zero, the curl in spherical coordinates becomes

$$\begin{aligned}\vec{\nabla} \times \vec{E} &= \frac{1}{r}\frac{1}{\sin\theta}\frac{\partial E_r}{\partial \phi}\hat{\theta} + \frac{1}{r}\left(-\frac{\partial E_r}{\partial \theta}\right)\hat{\phi} \\ &= \frac{1}{r}\frac{1}{\sin\theta}\frac{\partial}{\partial \phi}\left(\frac{k_e q}{r}\right)\hat{\theta} + \frac{1}{r}\left[-\frac{\partial}{\partial \theta}\left(\frac{k_e q}{r}\right)\right]\hat{\phi} \\ &= 0.\end{aligned}$$

This is not a surprising result in light of the radial nature of the electrostatic field of a point charge.

As mentioned in Chapter 2, vector fields with zero curl are called irrotational, and such fields have several important properties. One of those properties arises from the fact that the curl of a gradient is always zero: an irrotational vector field may always be written as the gradient of a scalar field.

In the case of electrostatic fields, the electric field may be written as the gradient of the scalar electric potential (usually written as ϕ or V). By convention, the electric field is written as the negative gradient of the scalar potential, so you're likely to see this relationship written as

$$\vec{E} = -\vec{\nabla} V, \tag{3.31}$$

where V is the scalar electric potential with units of Nm/C (equivalent to joules per coulomb or volts).

Since the electric field is the negative of the change in electric potential with distance, moving along an electric field line in the direction it's pointing means that you're moving toward a region of lower electric potential. Likewise, moving in the opposite direction (opposite to the direction of the field) takes you into a region of higher potential, and moving perpendicular to the field lines results in no change in potential. Hence the "equipotential" surfaces are always perpendicular to the electric field lines.

Another differential vector operation useful in electrostatics is the Laplacian (∇^2). Recall that the Laplacian involves the second spatial derivative, specifically the divergence of the gradient. Since the electrostatic field $\vec{E}$ may be written as the negative of the gradient of the scalar potential V, taking the divergence of the electric field gives:

$$\vec{\nabla} \circ \vec{E} = \vec{\nabla} \circ (-\vec{\nabla} V) = -\nabla^2 V. \tag{3.32}$$

Since Gauss's Law says that the divergence of the electrostatic field must equal ρ/ϵ_0, this means

$$\nabla^2 V = -\rho/\epsilon_0. \tag{3.33}$$

This is known as Poisson's Equation. Since the Laplacian finds peaks and valleys of a function (locations at which the value of the function differs from the average value at surrounding locations), Poisson's Equation tells you that the electric potential can have local maxima and minima only at locations at which charge is present (that is, where $\rho \neq 0$). And if you recall that the Laplacian is negative at peaks and positive at valleys, you can see that positive charge produces a peak in electric potential while negative charge produces a valley. This is one reason that the electric field is taken as the negative gradient of the electric potential.

In regions in which the electric charge density (ρ) is zero, Poisson's Equation becomes Laplace's Equation:

$$\nabla^2 V = 0, \tag{3.34}$$

so there are no maxima or minima in electric potential for locations with zero charge density.

3.4 The magnetic field

In this section, you can read about the behavior of the magnetic field ($\vec{B}$) and the magnetic force on a moving charged particle. You'll also find a discussion of the application of the vector operations of divergence and curl to the magnetostatic field.

Unlike electrostatic field lines, which diverge from positive charge and converge on negative charge, magnetic field lines form circles around the electric current (flowing charge) that is producing the magnetic field. And just as stationary source charges produce electrostatic fields, stationary currents (in which the charge flow is constant) produce magnetic fields that are called "magnetostatic." An example of such a field is shown in Figure 3.20. The direction of those field lines is determined using the right-hand rule: if you put the thumb of your right hand along the direction of current flow and curl

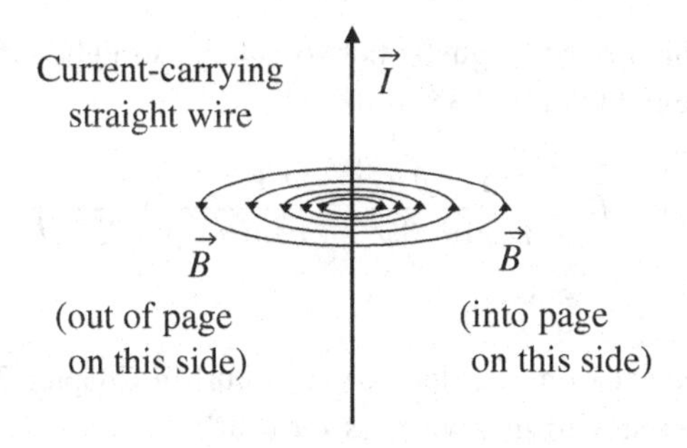

Figure 3.20 Magnetic field of a long, straight wire.

your fingers (like you're grabbing the current), the magnetic field points in the direction of your curled fingers. So if you were to reverse the direction of that current flow, the magnetic field lines would still form circles around the current, but the magnetic field lines would point in the opposite direction (as you can tell by observing the direction of your curled fingers when your thumb points in the opposite direction).

You can tell by the spacing of the field lines in Figure 3.20 that the strength of the magnetic field is decreasing as the distance from the current increases. For a thin wire of infinite length carrying current I, the vector magnetic field is given by the equation

$$\vec{B} = \frac{\mu_0 I}{2\pi r}\hat{\phi}, \tag{3.35}$$

where μ_0 is a constant called the magnetic permeability of free space, r is the distance from the wire to the point at which the magnetic field is being determined, and $\hat{\phi}$ is the cylindrical-coordinate unit vector that points in the direction circulating around the wire. The standard (SI) unit of magnetic field is the tesla (T).

Comparing the magnetic field lines around an electric current to the vector fields with various values of divergence and curl discussed in Chapter 2, you may have already guessed that magnetic fields fit into the "low divergence, high curl" category. Recall that electric field lines originate on positive charge and terminate on negative charge, and it is only at the location of those charges that the divergence of the electrostatic field is non-zero. And since magnetic field lines circulate back onto themselves rather than diverging from and converging upon specific locations, it's reasonable to expect small values for the divergence of the magnetic field. In fact, the divergence of the magnetic field ($\vec{B}$) is exactly zero, as indicated by Gauss's Law for magnetic fields:

$$\vec{\nabla} \circ \vec{B} = 0. \tag{3.36}$$

You can verify this for the magnetic field of a long, straight wire by taking the divergence of the field in Eq. 3.35:

$$\begin{aligned}\vec{\nabla} \circ \vec{B} &= \frac{1}{r\sin\theta}\frac{\partial B_\phi}{\partial \phi} = \frac{1}{r\sin\theta}\frac{\partial}{\partial \phi}\left(\frac{\mu_0 I}{2\pi r}\right)\\ &= 0.\end{aligned}$$

As you might expect from the discussion of curl in Chapter 2, the magnetic field around a current-carrying wire has zero curl:

$$\vec{\nabla} \times \vec{B} = \left(-\frac{\partial B_\phi}{\partial z}\right)\hat{r} + \frac{1}{r}\left(\frac{\partial (rB_\phi)}{\partial r}\right)\hat{z}$$
$$= \left[-\frac{\partial}{\partial z}\left(\frac{\mu_0 I}{2\pi r}\right)\right]\hat{r} + \frac{1}{r}\left[\frac{\partial}{\partial r}\left(r\frac{\mu_0 I}{2\pi r}\right)\right]\hat{z}$$
$$= 0.$$

As in the case of the divergence of the electric field, which has a non-zero value only at locations at which charge exists, the only locations at which the curl of the magnetic field is non-zero are locations at which current exists (that is, at the singularity point $r = 0$).

Other uses of vectors and vector operations come about when you consider the force ($\vec{F}_B$) produced by a magnetic field ($\vec{B}$) on a moving electric charge (q). This force is given by the vector equation

$$\vec{F}_B = q\vec{v} \times \vec{B}, \qquad (3.37)$$

where $\vec{v}$ is the velocity of the charged particle with respect to the magnetic field. The magnitude of the force is readily found using the definition of the magnitude of the vector cross product ($|\vec{A} \times \vec{B}| = |\vec{A}||\vec{B}| \sin\theta$):

$$|\vec{F}_B| = q|\vec{v}||\vec{B}| \sin\theta, \qquad (3.38)$$

where θ is the angle between vector $\vec{v}$ and vector $\vec{B}$.

Examined carefully, Eqs. 3.37 and 3.38 can tell you a great deal about how magnetic fields affect charged particles. Compare these equations to Eq. 3.27 ($\vec{F}_E = q\vec{E}$), and note that there are similarities and differences between electric and magnetic forces:

- Similarity: Both are directly proportional to the amount of charge (q);
- Similarity: Both are directly proportional to the field strength ($\vec{E}$ or $\vec{B}$);
- Difference: The velocity ($\vec{v}$) of the particle appears in the magnetic equation;
- Difference: The magnetic force depends on the angle between the velocity and the magnetic field;
- Difference: The magnetic force is perpendicular to both the velocity and the magnetic field.

The similarities seem reasonable: both electric and magnetic forces are stronger if the fields are stronger and if the amount of charge is greater. Also, charges with opposite signs feel forces in opposite directions. The first listed difference (the fact that the magnetic force depends on the *velocity* of the particle) has the interesting consequence that a charged particle at rest with respect to the magnetic field ($\vec{v} = 0$) feels no force whatsoever from that field. And for particles moving with respect to the magnetic field, the faster the particle moves, the stronger the magnetic force becomes.

The presence of the vector cross product in the magnetic force equation also has some important consequences. One of those consequences is that charged particles moving in a direction parallel or antiparallel to the magnetic field feel zero magnetic force. That's because in both the parallel ($\theta = 0°$) and antiparallel ($\theta = 180°$) cases, the sine term in Eq. 3.38 is zero. So the closer the angle θ between $\vec{v}$ and $\vec{B}$ is to 90°, the stronger the magnetic force.

Another consequence of the vector cross product in Eq. 3.37 is that the magnetic force ($\vec{F}_B$) can never point in the direction of the magnetic field, since the vector result of the cross product is by definition perpendicular to both vectors forming the product ($\vec{v}$ and $\vec{B}$ in this case). For this same reason, the magnetic force can never point in the direction of the particle's velocity vector, and must in fact be perpendicular to that vector. So if you imagine the flat plane formed by the velocity vector and the magnetic field, you can be sure that the magnetic force (if any) must be perpendicular to that plane.

If you've read the discussion of radial and tangential acceleration in Section 3.2, you should understand that this means that magnetic fields can provide radial but never tangential acceleration to a charged particle (since tangential acceleration requires a component of force that's either parallel or antiparallel to the velocity vector). And since $\vec{v} \times \vec{B}$ always points perpendicular to $\vec{v}$, magnetic fields can provide only radial acceleration. Thus magnetic fields may change the direction but never the speed of charged particles.

An example of the geometry involved in magnetic force is shown in Figure 3.21. In this figure, the direction of the magnetic field is into the page, as

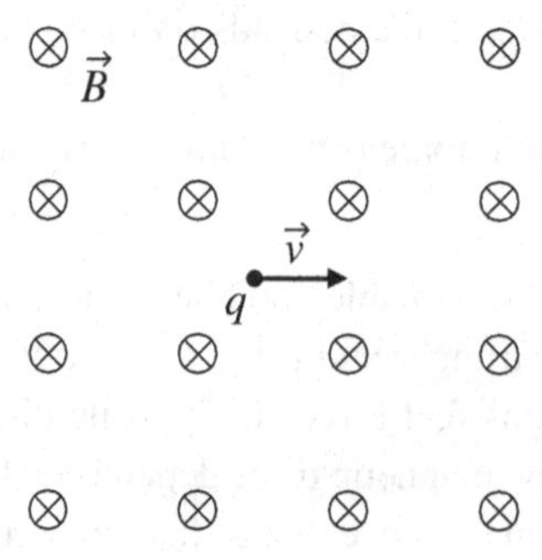

Figure 3.21 Charged particle moving to right; magnetic field into page.

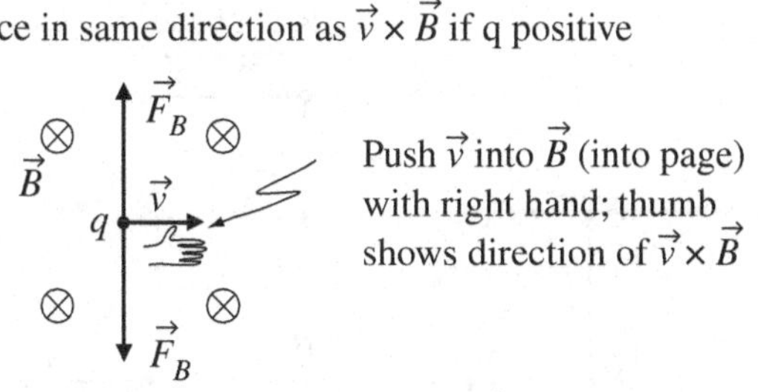

Figure 3.22 Magnetic force for positive and negative charges.

indicated by the crosses inside circles,[11] and the charged particle (q) is moving to the right.

To determine the direction of the magnetic force in this case, you simply have to imagine forming the vector cross product $\vec{v} \times \vec{B}$ using the right-hand rule, as shown in Figure 3.22. Once you know the direction of $\vec{v} \times \vec{B}$, it's very important to remember (but easy to forget) that you must then reverse the direction if the charge q is negative (since by Eq. 3.37, $\vec{F_B} = q\vec{v} \times \vec{B}$, meaning that the magnetic force is *opposite* to the direction of $\vec{v} \times \vec{B}$ if q is negative). This explains why two directions for the magnetic force $\vec{F_B}$ are shown in Figure 3.22: upward if q is positive and downward if q is negative.

Once you understand the direction of the magnetic force relative to the velocity of the charged particle, it should help explain why you may have heard or read about charged particles "circling around magnetic field lines" or perhaps "spiralling along the magnetic field." Consider the positively charged particle q in Figure 3.23. If this particle is initially at the leftmost position in the figure, travelling with velocity $\vec{v}$ straight up the page, and the magnetic field $\vec{B}$ points directly *out* of the page, the direction of the magnetic force $q\vec{v} \times \vec{B}$ is initally to the right (as you can determine using the right-hand rule). This force causes the particle to travel on the dashed path to the topmost position in the figure. At that point, the magnetic force $\vec{F_B}$ points straight down the page. Just as at the previous position, since q is positively charged, the magnetic force points in the same direction as $\vec{v} \times \vec{B}$. This now-downward force causes the particle to travel to the rightmost position, at which point the velocity is straight

[11] This is common notation in physics and engineering; you can remember it by thinking of a hunter's feathered arrow. Seen from the back, you can see the back edges of the feathers, so it looks like this: ⊗. But seen from the front, you can see the arrow's point, so it looks like this: ⊙.

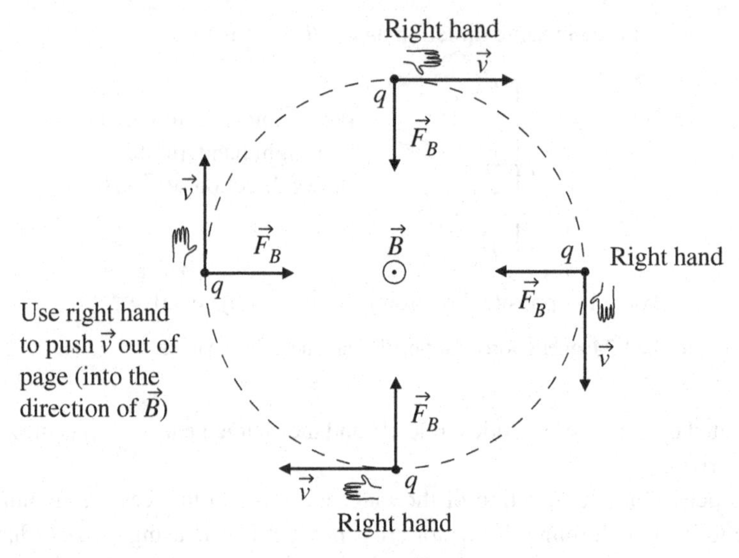

Figure 3.23 Magnetic force on positive charge.

down the page and the magnetic force $\vec{F}_B$ points directly to the left. This force causes the particle to reach the bottom position in Figure 3.23, at which point the velocity is to the left and the magnetic force points straight up the page. Under the influence of this force, the particle will travel back to the starting (leftmost) position, and the entire cycle will repeat. So this positively charged particle makes a clockwise circle around the outward-pointing magnetic field.

Applying the same reasoning to a negatively charged particle, you should be able to determine that it will make counter-clockwise circles around the same outward-pointing magnetic field. And if the field direction is reversed, so that $\vec{B}$ points into the page rather than outward, the sense of the particle's rotation will be reversed (so that a positively charged particle will circle counter-clockwise and a negatively charged particle will circle in the clockwise direction).

The particles in these examples retrace the same path over and over, so what makes some particles "spiral around" the lines of the magnetic field? Simply this: the particle's velocity must have a component parallel (or antiparallel) to the direction of the magnetic field. Note that the particle shown in Figure 3.23 is moving entirely in the plane of the page, and the magnetic field is perpendicular to the page. Hence the particle's velocity vector has no component along the magnetic field (into or out of the page). If such a component were present, the particle would have a component of its motion along the field lines while also circling around them. In that case, the circular path shown in Figure 3.23 would move into or out of the paper over time, and the circle would become a spiral. The magnetic field has no effect on the velocity component ($v_{||}$) parallel or antiparallel to the field (since there's no magnetic force in that direction), so the speed with which the particle moves along the field line is constant as long as no other forces are acting.

3.5 Chapter 3 problems

3.1 Solve the box-on-a-ramp problem (that is, find the acceleration of the box) for the frictionless case using a Cartesian coordinate system for which the y-axis points vertically upward and the x-axis points horizontally to the right.

3.2 The maximum force of static friction is $\mu_s \vec{F}_n$, where μ_s is the coefficient of static friction and $\vec{F}_n$ is the normal force. How big must the coefficient of static friction μ_s be to prevent a box of mass m from sliding down a ramp inclined 20 degrees from the horizontal?

3.3 If a delivery woman pushes a box of mass m up a 2 m ramp with a force of 10 N, how fast is the box moving at the top of the ramp if the ramp angle to the horizontal is 25 degrees and the coefficient of kinetic friction is 0.33?

3.4 If the hammer-thrower shown on the cover of this book wishes to launch a hammer of mass 7.26 kg on a cable of length 1.22 m with a speed of 22 m/s, what is the magnitude of the centripetal force he must supply?

3.5 Imagine a Formula 1 car going around a curve with radius of 10 m while slowing from a speed of 180 mph to 120 mph in 2 s. What are the magnitude and direction of the car's acceleration at the instant the car's speed is 150 mph?

3.6 If three electric charges q_1, q_2, and q_3 have the values and locations shown in Figure 3.18, find the electric field they produce at the origin ($x = 0$, $y = 0$), then use your value of the field to determine the electric force on an electron at that location.

3.7 If the vector electric field $\vec{E}$ in some region is given in spherical coordinates by $\frac{5}{r}\hat{r} + \frac{2}{r}\sin\theta\cos\phi\,\hat{\theta} - \frac{1}{r}\sin\theta\cos\phi\,\hat{\phi}$ (N/C), what is the volume charge density ρ in that region?

3.8 If the scalar electric potential V in some region is given in cylindrical coordinates by $V(r, \phi, z) = r^2 sin\phi\, e^{-3/z}$, what is the electric field $\vec{E}$ in that region?

3.9 For the scalar electric potential V of Problem 3.8, use Poisson's Equation to find volume charge density ρ in that region.

3.10 Find the magnitude and direction of the magnetic force on a charged particle with charge -4 nC and velocity $\vec{v} = 2.5 \times 10^4\,\hat{\imath} + 1.1 \times 10^4\,\hat{\jmath}$ (m/s) if the magnetic field in the region is given by $\vec{B} = 1.2 \times 10^{-3}\,\hat{\imath} + 5.6 \times 10^{-3}\,\hat{\jmath} - 3.2 \times 10^{-3}\,\hat{k}$ (T).

4
Covariant and contravariant vector components

The vector concepts and techniques described in the previous chapters are important for two reasons: they allow you to solve a wide range of problems in physics and engineering, and they provide a foundation on which you can build an understanding of tensors (the "facts of the universe"). To achieve that understanding, you'll have to move beyond the simple definition of vectors as objects with magnitude and direction. Instead, you'll have to think of vectors as objects with components that transform between coordinate systems in specific and predictable ways. It's also important for you to realize that vectors can have more than one kind of component, and that those different types of component are defined by their behavior under coordinate transformations.

So this chapter is largely about the different types of vector component, and those components will be a lot easier to understand if you have a solid foundation in the mathematics of coordinate-system transformation.

4.1 Coordinate-system transformations

In taking the step from vectors to tensors, a good place to begin is to consider this question: "What happens to a vector when you change the coordinate system in which you're representing that vector?" The short answer is that nothing at all happens to the vector itself, but the vector's components may be different in the new coordinate system. The purpose of this section is to help you understand how those components change.

Before getting to that, you should spend a few minutes considering the statement that the vector itself doesn't change if you change the coordinate system. This may seem obvious in the case of scalars – after all, whether you measure temperature in Celsius or Fahrenheit doesn't make a room feel hotter or colder. Now remember that vectors are mathematical representations of physical entities, and those entities don't change just because you change the coordinate system in which you're representing them. Think about it: does the size of a room change if you tilt your head to one side? Clearly not. But if you use your tilted head to define up and down, then the points you designate as the top and bottom of the room may change, and this will change what you call the "height" and "width" of the room. The important idea is that the room itself doesn't change (it "remains invariant") under such a change of coordinate system. And if you define the center of your head to be the origin of your coordinate system, then walking toward one wall will "offset" the room (that is, the x, y, and z values of locations within the room may change), but once again the room itself is unchanged. Likewise, specifying dimensions of the room in inches rather than meters will allow you to put larger numbers in the real-estate ad, but that doesn't mean your room will hold a bigger sofa.

So if coordinate-system transformations such as rotation, translation, and scaling leave physical quantities unchanged, what exactly does happen to a vector when you transform coordinates? To understand that, consider the simple rotation of the two-dimensional Cartesian coordinate system shown in Figure 4.1. In this transformation, the location of the origin has not changed, but both the x- and y-axis have been tilted counter-clockwise by an angle θ. The rotated axes are labeled x' and y' and are drawn using dashed lines to distinguish them from the original axes.

What impact does this rotation have on a vector in this space? Take a look at vector $\vec{A}$ and its components in Figure 4.2(a) and (b). Note that the rotation has no effect on the length or direction of $\vec{A}$ (at first glance, $\vec{A}$ may look a bit different in Figure 4.2(a) and 4.2(b), but you can verify using a ruler and protractor that the vector itself is exactly the same). But the rotation has clearly caused the components of $\vec{A}$ to change: A'_x (the x'-component of A in the tilted coordinate system) is longer than A_x, and A'_y is shorter than A_y. If you

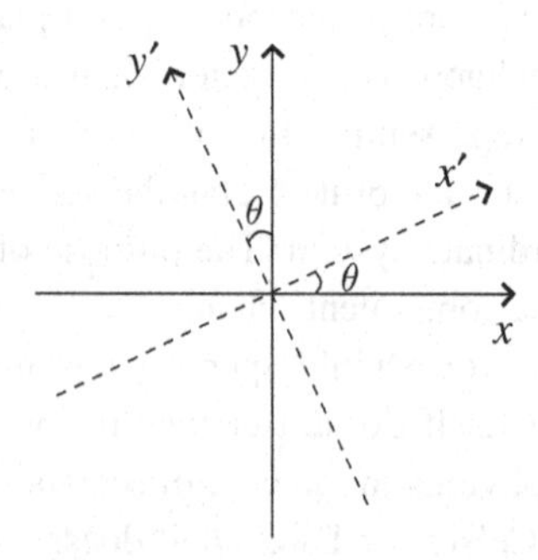

Figure 4.1 Rotation of 2-D coordinate system.

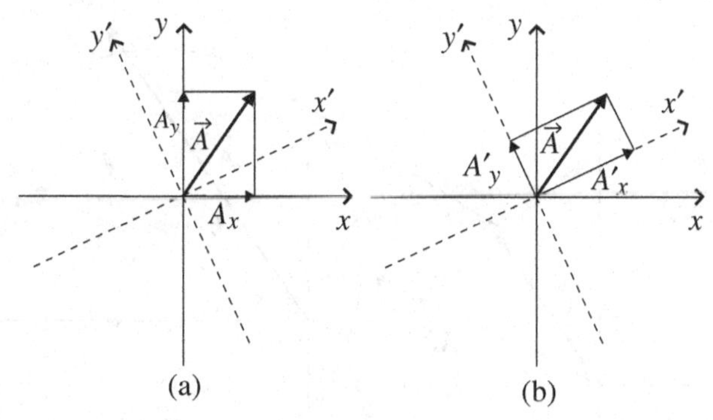

Figure 4.2 Change in vector components due to rotation of coordinate system.

were to continue rotating your axes in the same direction, you'd eventually reach an angle at which $\vec{A}$ lies entirely along the x'-axis, at which point the y'-component of $\vec{A}$ would vanish (that is, $A'_y = 0$) and the x'-component would equal the length of $\vec{A}$ ($A'_x = |\vec{A}|$).

Finding the change in the components of a vector due to rotation of the coordinate axes can be done both graphically using simple geometry and analytically using the dot product. You'll find the graphical approach in this section; the analytical approach is the subject of one of the problems at the end of this chapter.

If you think about the changes to A_x and A_y in Figure 4.2, you might come to realize that the vector component A'_x in the rotated coordinate system cannot depend entirely on the component A_x in the original system. After all, A_x contains some but not all of the information about vector $\vec{A}$; the rest is in A_y. And as the axes rotate, the axis that had pointed exclusively in the x-direction now points partially in the (former) y-direction. So it seems reasonable that the portion of $\vec{A}$ that had previously pointed in the original y-direction (and so contributed only to A_y) now points partially in the x'-direction, and hence contributes to the x'-component as well as the y'-component.

You can see how this works in Figure 4.3. The (a) portion of this figure shows how the vector component A_x in the original (non-rotated) coordinate system contributes to A'_x in the rotated system, and the (b) portion shows how the vector component A_y in the original system contributes to A'_x in the rotated system.

As you can see in both portions of the figure, A'_x can be considered to be made up of two segments, labeled ℓ_1 and ℓ_2. So

$$A'_x = \ell_1 + \ell_2, \tag{4.1}$$

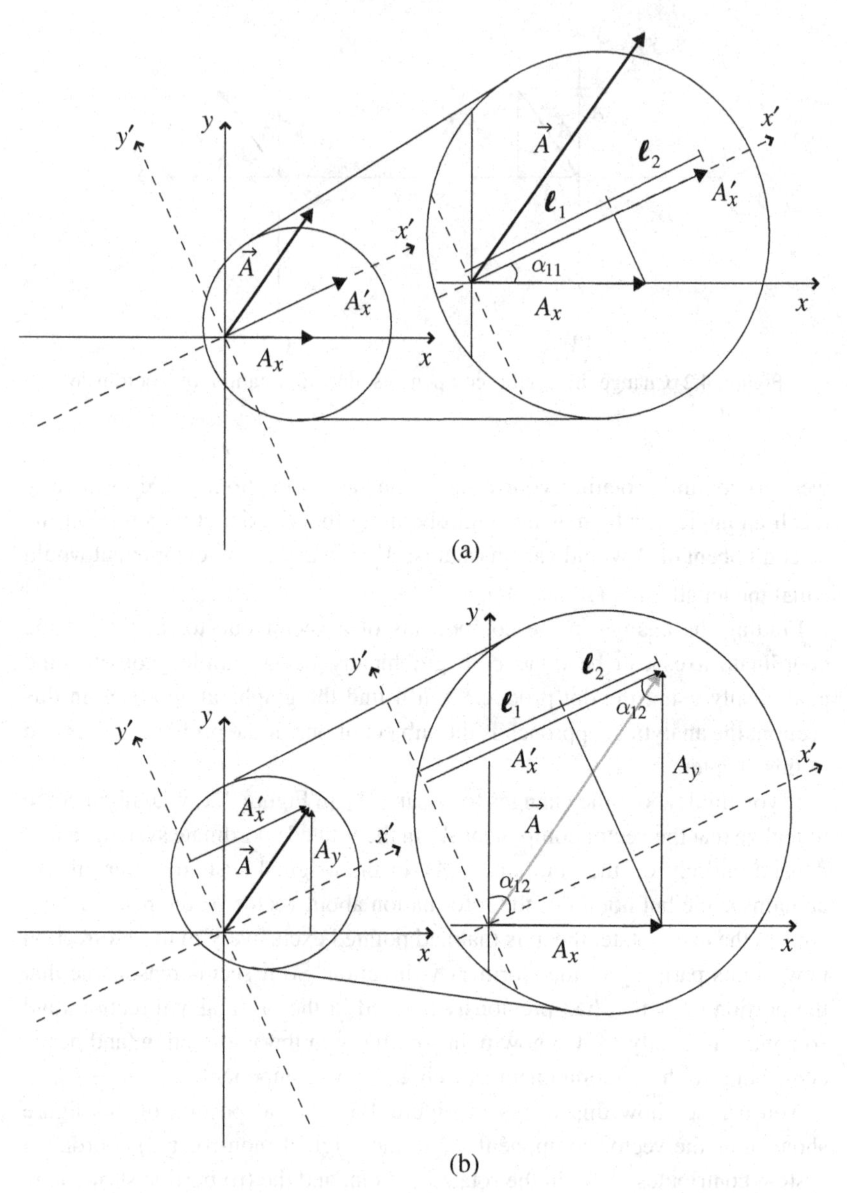

Figure 4.3 Dependence of A'_x on A_x and A_y.

and to determine how these segments depend on A_x and A_y, consider the right triangles shown in Figure 4.3. In the (a) portion of the figure, you can see that A_x is the hypotenuse of a right triangle formed by drawing a perpendicular from the end of A_x to the x'-axis. Call the angle between the x-axis and the x'-axis α_{11} (the reason for using double subscripts will become clear when rotations are written in matrix notation). Then the length of ℓ_1 (the projection of A_x onto the x'-axis) is $A_x \cos(\alpha_{11})$. Hence

$$\ell_1 = A_x \cos(\alpha_{11}). \tag{4.2}$$

To find the length of ℓ_2, consider the right triangle shown in Figure 4.3(b). In this case, the triangle is formed by sliding A'_x upward along the y'-axis and then drawing a perpendicular from the tip of A'_x to the x-axis. From this triangle, you should be able to see that

$$\ell_2 = A_y \cos(\alpha_{12}), \tag{4.3}$$

where α_{12} is the angle formed by the tips of A'_x and A_y (which is also the angle between the x'-axis and the y-axis, as you can see from the parallelogram in Figure 4.3(b).

Adding the expressions for ℓ_1 and ℓ_2, you can write A'_x as

$$A'_x = A_x \cos(\alpha_{11}) + A_y \cos(\alpha_{12}), \tag{4.4}$$

where A_x and A_y are the components of vector $\vec{A}$ in the non-rotated coordinate system, α_{11} is the angle between the x'-axis and the x-axis, and α_{12} is the angle between the x'-axis and the y-axis. You should note that the new component (A'_x) is a weighted linear combination of the original components (A_x and A_y). "Weighted" because the cosine factors determine how heavily each of the original components contributes to the new one, "linear" because the original components appear to the first power only, and "combination" because both A_x and A_y contribute to A'_x.

A similar analysis for A'_y, the y-component of vector $\vec{A}$ in the rotated coordinate system, gives

$$A'_y = A_x \cos(\alpha_{21}) + A_y \cos(\alpha_{22}), \tag{4.5}$$

where α_{21} is the angle between the y'-axis and the x-axis, and α_{22} is the angle between the y'-axis and the y-axis.

The relationship between the components of vector $\vec{A}$ in the rotated and non-rotated systems is conveniently expressed using vector/matrix notation[1] as

[1] Remember, there's a review of matrix notation and algebra on the book's website.

$$\begin{pmatrix} A'_x \\ A'_y \end{pmatrix} = \begin{pmatrix} \cos(\alpha_{11}) & \cos(\alpha_{12}) \\ \cos(\alpha_{21}) & \cos(\alpha_{22}) \end{pmatrix} \begin{pmatrix} A_x \\ A_y \end{pmatrix}. \tag{4.6}$$

This is called a "transformation equation" for the components of vector $\vec{A}$, and the two-column matrix is called a "transformation matrix." The elements of that matrix are called the "direction cosines." Note that for a rigid rotation of the Cartesian axes through angle θ, the angles α_{11} and α_{22} are both equal to θ, while $\alpha_{12} = 90° - \theta$ and $\alpha_{21} = 90° + \theta$. The transformation matrix in this case is

$$\begin{pmatrix} \cos(\theta) & \cos(90° - \theta) \\ \cos(90° + \theta) & \cos(\theta) \end{pmatrix} = \begin{pmatrix} \cos(\theta) & \sin(\theta) \\ -\sin(\theta) & \cos(\theta) \end{pmatrix}, \tag{4.7}$$

since $\cos(90° - \theta) = \sin(\theta)$ and $\cos(90° + \theta) = -\sin(\theta)$.

To understand how this works in practice, consider vector $\vec{A}$ given as

$$\vec{A} = 5\hat{\imath} + 3\hat{\jmath} \tag{4.8}$$

in a two-dimensional Cartesian coordinate system. Now imagine that the x- and y-axes of that coordinate system are rotated counter-clockwise by 150°, as shown in Figure 4.4.

Before jumping to the equations to find the components A'_x and A'_y in the rotated coordinate system, it's worth a few minutes to take a look at the diagram to estimate what the effect of the rotation on the components will be. From Figure 4.4(b), it's pretty clear that both the A'_x and A'_y components will be negative, and the A'_y component appears to be somewhat larger than the A'_x component.

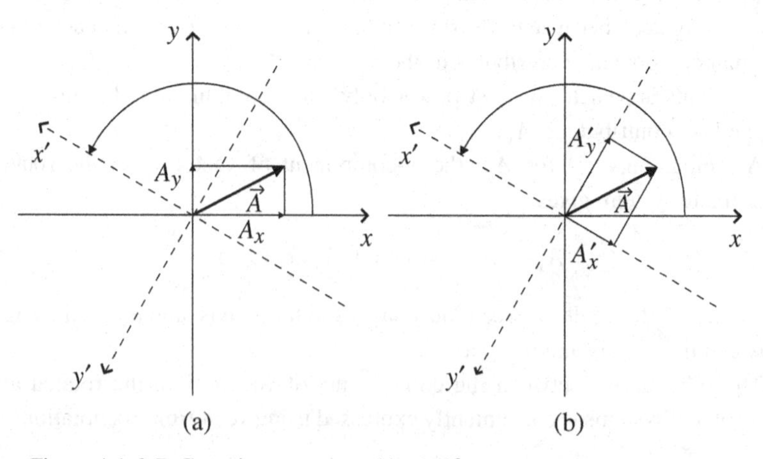

Figure 4.4 2-D Cartesian axes rotated by 150°.

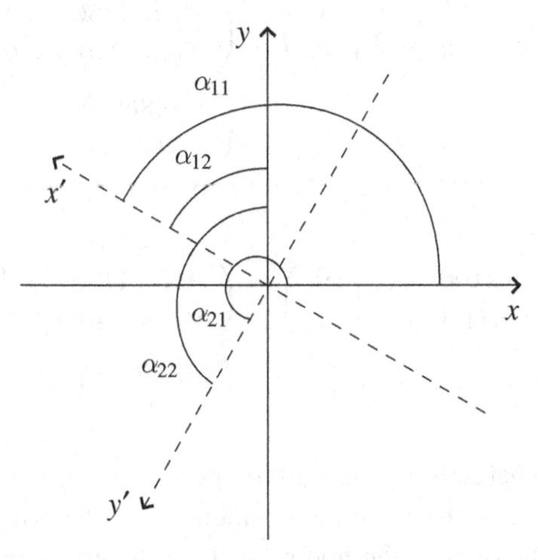

Figure 4.5 Angles between original and rotated axes.

Now that you have an idea of what to expect, you can insert the relevant values into Eq. 4.6. You know that $A_x = 5$ and $A_y = 3$, and using the angles shown in Figure 4.5, you should be able to see that $\alpha_{11} = 150°$, $\alpha_{12} = 60°$, $\alpha_{21} = 240°$, and $\alpha_{22} = 150°$.

So you have

$$\begin{pmatrix} A'_x \\ A'_y \end{pmatrix} = \begin{pmatrix} \cos(150°) & \cos(60°) \\ \cos(240°) & \cos(150°) \end{pmatrix} \begin{pmatrix} A_x \\ A_y \end{pmatrix}, \tag{4.9}$$

or

$$A'_x = 5\cos(150°) + 3\cos(60°) = -2.8, \tag{4.10}$$

and

$$A'_y = 5\cos(240°) + 3\cos(150°) = -5.1. \tag{4.11}$$

As a quick visual analysis suggested, both components are negative and the y'-component is larger than the x'-component in the rotated system.

It is very important for you to understand that the transformation equation (4.6) does not rotate or change the vector $\vec{A}$ in any way; it determines the values of the components of vector $\vec{A}$ in a new coordinate system. This distinction is important because you may be tempted to apply this transformation matrix to basis vectors such as $\hat{\imath}$ (1, 0) and $\hat{\jmath}$ (0, 1), which for a counter-clockwise 150° rotation gives for $\hat{\imath}$

$$\begin{pmatrix} \cos(150°) & \cos(60°) \\ \cos(240°) & \cos(150°) \end{pmatrix} \begin{pmatrix} 1 \\ 0 \end{pmatrix} = \begin{pmatrix} 1\cos(150°) + 0\cos(60°) \\ 1\cos(240°) + 0\cos(150°) \end{pmatrix}$$
$$= \begin{pmatrix} -0.866 \\ -0.5 \end{pmatrix}, \tag{4.12}$$

and for $\hat{\jmath}$

$$\begin{pmatrix} \cos(150°) & \cos(60°) \\ \cos(240°) & \cos(150°) \end{pmatrix} \begin{pmatrix} 0 \\ 1 \end{pmatrix} = \begin{pmatrix} 0\cos(150°) + 1\cos(60°) \\ 0\cos(240°) + 1\cos(150°) \end{pmatrix}$$
$$= \begin{pmatrix} 0.5 \\ -0.866 \end{pmatrix}. \tag{4.13}$$

There's nothing inherently wrong with doing this, as long as you remember what the results mean: these are the components of the original unit vectors $\hat{\imath}$ and $\hat{\jmath}$ (that is, the ones in the non-rotated coordinate system) expressed in terms of the rotated coordinate axes, as you can see in Figure 4.6. These are *not* the unit vectors $\hat{\imath}'$ and $\hat{\jmath}'$ which point in the direction of the x' and y'-axes (remember that in the primed coordinate system, the unit vectors $\hat{\imath}'$ and $\hat{\jmath}'$, pointing along the rotated coordinate axes, must have components (1, 0) and (0, 1), respectively).

Rigid rotation of Cartesian axes is only one type of the myriad coordinate transformations that can change the components of a vector. But as long as the new components can be written as weighted sums of the original components, the transformation is linear and can be represented by a matrix equation. For

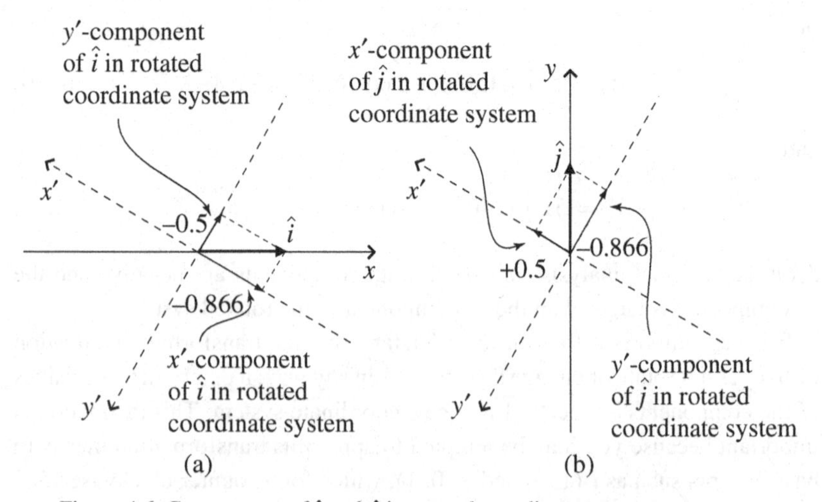

Figure 4.6 Components of $\hat{\imath}$ and $\hat{\jmath}$ in rotated coordinate system.

reasons that will become clear when you read Section 4.3 of this chapter, such transformations of vector components are called "inverse" or "passive" transformations, which means the matrix equation of such a transformation will look like this:

$$\begin{pmatrix} \text{Components of} \\ \text{same vector} \\ \text{in new system} \end{pmatrix} = \begin{pmatrix} \text{Inverse} \\ \text{transformation} \\ \text{matrix} \end{pmatrix} \begin{pmatrix} \text{Components of} \\ \text{vector in} \\ \text{original system} \end{pmatrix}. \tag{4.14}$$

At this point, you may be wondering how you might go about transforming the unit vectors of the original (non-rotated) system (that is, $\hat{\imath}$ and $\hat{\jmath}$) into the unit vectors of the primed (rotated) system ($\hat{\imath}'$ and $\hat{\jmath}'$). That's a different question, because you're no longer asking, "Given the components of a vector in one coordinate system, how do I find the components of that same vector in a different coordinate system?" Instead, you're asking, "How do I change a given vector (in this case, a unit vector in one coordinate system) into a different vector (the unit vector in a different coordinate system)?" That question is addressed in the next section.

4.2 Basis-vector transformations

The previous section illustrated what happens to the components of a vector when the two-dimensional Cartesian axes are rotated, and the results are not surprising: the components of the vector referenced to the new (rotated) axes are different from the components referenced to the original (non-rotated) axes. More specifically, the new components are weighted linear combinations of the original components.

Now here's a very important point: as your studies carry you along the path from vectors to tensors, you will undoubtedly run across discussions of "covariant" and "contravariant" vector components.[2] In those discussions, you may see words to the effect that covariant components transform in the same way as basis vectors ("co" ≈ "with"), and contravariant components transform in the opposite way to basis vectors ("contra" ≈ "against"). As you'll see later in this chapter, there's plenty of truth in that description, but there's also a major pitfall. That's because the "transformation" of basis vectors usually refers to the conversion of the basis vectors in the original (non-rotated) coordinate system to the different basis vectors which point along the coordinate axes in the new (rotated) system, whereas the "transformation" of vector

[2] These components are identical in the Cartesian coordinate systems considered so far.

components refers to the change in the components of the same vector referred to two different sets of coordinate axes. The potential for confusion here is sufficiently great to cause Schutz to write that "the reason that 'co' and 'contra' have been abandoned is that they mix up two very different things."[3] Schutz wrote that in 1983, and for better or worse, the "covariant/contravariant" terminology is still with us – that's why in this book you'll find those words as well as more modern terminology.

Why did the "covariant/contravariant" terminology take hold in the first place? Probably because the process of changing a vector into a different vector has much in common with the process of transforming the components of a vector from one coordinate system to another. This section shows you how to make a new vector using rotation (specifically, how to rotate basis vectors).

To understand the process of rotating a vector, consider vector $\vec{A}$ in Figure 4.7(a). The rotation shown in Figure 4.7(b) causes vector $\vec{A}$ to point in a different direction, which means it is no longer the same vector (which is why it's labeled $\vec{A}'$ after the rotation). The relationship between the components of the original (non-rotated) vector and the new (rotated) vector can be found rather easily through geometric constructions such as those shown in Figure 4.8. In this example, the rotation angle is α. The x- and y-components of vectors $\vec{A}$ and $\vec{A}'$ are

$$\begin{aligned} A_x &= |\vec{A}|\cos(\theta), \quad A'_x = |\vec{A}'|\cos(\theta'), \\ A_y &= |\vec{A}|\sin(\theta), \quad A'_y = |\vec{A}'|\sin(\theta'). \end{aligned}$$

But $\theta' = \alpha + \theta$, so the components A'_x and A'_y are

$$\begin{aligned} A'_x &= |\vec{A}'|\cos(\alpha + \theta) = |\vec{A}'|\left[\cos(\alpha)\cos(\theta) - \sin(\alpha)\sin(\theta)\right], \\ A'_y &= |\vec{A}'|\sin(\alpha + \theta) = |\vec{A}'|\left[\sin(\alpha)\cos(\theta) + \cos(\alpha)\sin(\theta)\right]. \end{aligned}$$

Since the length of $\vec{A}$ must be the same as the length of $\vec{A}'$ (the vector rotated but did not change length), you can write $|\vec{A}| = |\vec{A}'|$, which means that

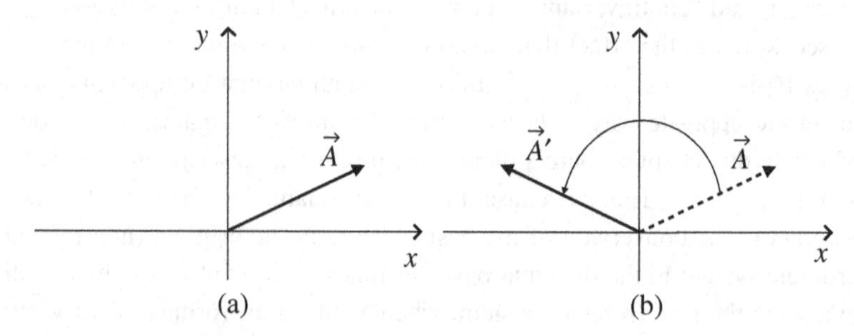

Figure 4.7 Rotation of a vector.

[3] Schutz, B., *A First Course in General Relativity*, p. 64. See further reading.

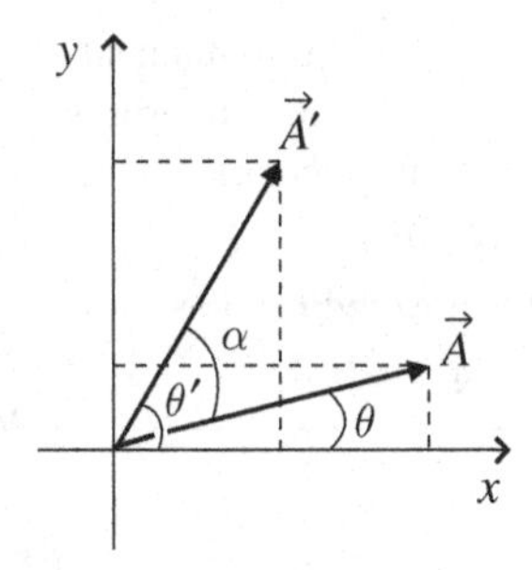

Figure 4.8 Angles involved in the rotation of a vector.

$$\begin{aligned} A'_x &= |\vec{A}'| \left[\cos(\alpha)\cos(\theta) - \sin(\alpha)\sin(\theta)\right] \\ &= |\vec{A}|\cos(\alpha)\cos(\theta) - |\vec{A}|\sin(\alpha)\sin(\theta), \\ A'_y &= |\vec{A}'| \left[\sin(\alpha)\cos(\theta) + \cos(\alpha)\sin(\theta)\right] \\ &= |\vec{A}|\sin(\alpha)\cos(\theta) + |\vec{A}|\cos(\alpha)\sin(\theta). \end{aligned}$$

But $|\vec{A}|\cos(\theta)$ is just A_x and $|\vec{A}|\sin(\theta)$ is A_y, so you can write

$$\begin{aligned} A'_x &= A_x\cos(\alpha) - A_y\sin(\alpha), \\ A'_y &= A_x\sin(\alpha) + A_y\cos(\alpha), \end{aligned}$$

or, as a matrix equation,

$$\begin{pmatrix} A'_x \\ A'_y \end{pmatrix} = \begin{pmatrix} \cos(\alpha) & -\sin(\alpha) \\ \sin(\alpha) & \cos(\alpha) \end{pmatrix} \begin{pmatrix} A_x \\ A_y \end{pmatrix}, \tag{4.15}$$

which tells you how to find the components A'_x and A'_y of the new vector ($\vec{A}'$) in the original coordinate system.

To see how this works in practice, consider a rotation such as the one shown in Figure 4.7, but through a larger rotation angle of $\alpha = 150°$. If the original vector is given by $\vec{A} = A_x\hat{\imath} + A_y\hat{\jmath} = 5\hat{\imath} + 3\hat{\jmath}$, then

$$\begin{pmatrix} A'_x \\ A'_y \end{pmatrix} = \begin{pmatrix} \cos(150°) & -\sin(150°) \\ \sin(150°) & \cos(150°) \end{pmatrix} \begin{pmatrix} 5 \\ 3 \end{pmatrix} = \begin{pmatrix} -5.83 \\ -0.10 \end{pmatrix}, \tag{4.16}$$

so the new vector $\vec{A}' = -5.83\hat{\imath} - 0.10\hat{\jmath}$. This means that by rotating vector $\vec{A}$ through 150°, you've produced a new vector that lies almost entirely along the negative x-axis (you can see this by noting that the x-component is negative and much larger than the y-component). Remember that this is a new vector expressed using the same basis ($\hat{\imath}$ and $\hat{\jmath}$) and is not the same vector expressed using a new basis (because in this case you rotated the vector, not the coordinate system).

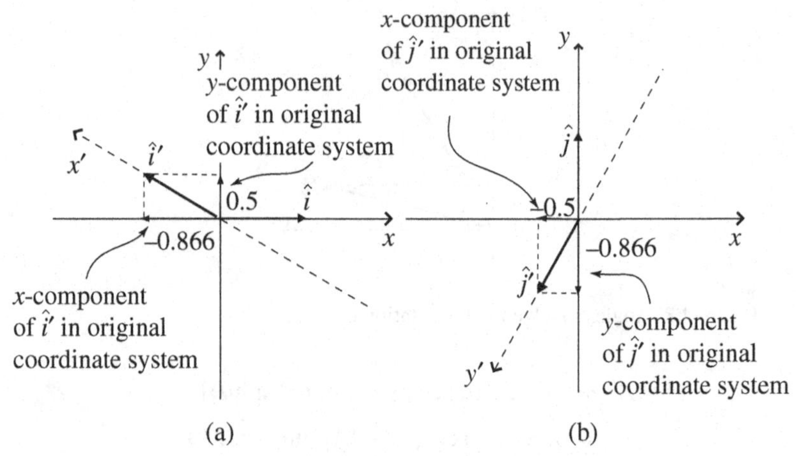

Figure 4.9 Components of $\hat{\imath}'$ and $\hat{\jmath}'$ in original (unrotated) coordinate system.

You can, of course, rotate the basis vectors $\hat{\imath}$ and $\hat{\jmath}$ using this same approach. This can be helpful if you're faced with a problem involving a rotated coordinate system and you wish to express the basis vectors pointing along the axes of the rotated system in terms of the basis vectors in the original (non-rotated) system. For example, to rotate the $\hat{\imath}$ unit vector by 150° counter-clockwise, you can use

$$\begin{pmatrix} \hat{\imath}'_x \\ \hat{\imath}'_y \end{pmatrix} = \begin{pmatrix} \cos(150°) & -\sin(150°) \\ \sin(150°) & \cos(150°) \end{pmatrix} \begin{pmatrix} 1 \\ 0 \end{pmatrix} = \begin{pmatrix} -0.866 \\ 0.5 \end{pmatrix}, \tag{4.17}$$

where $\hat{\imath}'_x$ represents the x-component of the 150°-rotated $\hat{\imath}$ vector and $\hat{\imath}'_y$ represents the y-component of the rotated $\hat{\imath}$ vector, as shown in Figure 4.9(a). You can also rotate the $\hat{\jmath}$ unit vector by the same angle using

$$\begin{pmatrix} \hat{\jmath}'_x \\ \hat{\jmath}'_y \end{pmatrix} = \begin{pmatrix} \cos(150°) & -\sin(150°) \\ \sin(150°) & \cos(150°) \end{pmatrix} \begin{pmatrix} 0 \\ 1 \end{pmatrix} = \begin{pmatrix} -0.5 \\ -0.866 \end{pmatrix}, \tag{4.18}$$

where $\hat{\jmath}'_x$ represents the x-component of the 150°-rotated $\hat{\jmath}$ vector and $\hat{\jmath}'_y$ represents the y-component of the rotated $\hat{\jmath}$ vector, as shown in Figure 4.9(b).

Just as in Eq. 4.15, the new components of the $\hat{\imath}'$ and $\hat{\jmath}'$ vectors are expressed in the same coordinate system as the original $\hat{\imath}$ and $\hat{\jmath}$. As pointed out in the previous section, the components of $\hat{\imath}'$ and $\hat{\jmath}'$ in the rotated coordinate system must be $(1, 0)$ and $(0, 1)$.

So if you wish to transform a set of basis vectors into new basis vectors (pointing along different coordinate axes), you use a "direct" or "active" transformation matrix, and the matrix equation looks like this:

$$\begin{pmatrix} \text{New basis} \\ \text{vectors} \end{pmatrix} = \begin{pmatrix} \text{Direct} \\ \text{transformation} \\ \text{matrix} \end{pmatrix} \begin{pmatrix} \text{Original basis} \\ \text{vectors} \end{pmatrix}. \tag{4.19}$$

Comparing this to Eq. 4.14 should help you understand that transformation matrices can be used for two different but related operations: finding the components of the same vector in a new coordinate system or finding the components of a different vector (such as a new basis vector) in the original coordinate system. The next section presents a comparison of these two types of transformation matrix.

4.3 Basis-vector vs. component transformations

Since Eq. 4.14 and Eq. 4.19 both involve transformation matrices, it's natural to wonder how those transformation matrices might be related. You can find a clue to that relationship by comparing the transformation matrix in Eq. 4.7 (pertaining to component change due to a coordinate-axis rotation through angle θ) with that of Eq. 4.15 (pertaining to basis-vector rotation through angle θ). Extracting the transformation matrix from each of those equations gives:

From Eq. 4.7:

$$\begin{pmatrix} \cos(\theta) & \sin(\theta) \\ -\sin(\theta) & \cos(\theta) \end{pmatrix}$$

↖

Transformation matrix for finding components of same vector as coordinate system is rotated through angle θ

From Eq. 4.15:

$$\begin{pmatrix} \cos(\theta) & -\sin(\theta) \\ \sin(\theta) & \cos(\theta) \end{pmatrix}$$

↖

Transformation matrix for finding new basis vectors by rotating original basis vectors through angle θ

Multiplying these two matrices reveals the nature of the relationship between them:

$$\begin{pmatrix} \cos(\theta) & \sin(\theta) \\ -\sin(\theta) & \cos(\theta) \end{pmatrix} \begin{pmatrix} \cos(\theta) & -\sin(\theta) \\ \sin(\theta) & \cos(\theta) \end{pmatrix} = \begin{pmatrix} 1 & 0 \\ 0 & 1 \end{pmatrix}.$$

This means that in this case the component-transformation matrix is the inverse of the basis-vector transformation matrix (since multiplying a matrix by its inverse produces the identity matrix). The fact that in this case the transpose of the transformation matrix is equal to its inverse means that this transformation matrix is "orthogonal" (converting from one Cartesian coordinate system into a different one).

In light of the inverse relationship between the basis-vector transformation matrix and the vector-component transformation matrix, you might say that in this case the vector components transform inversely to or "against" the manner in which the basis vectors transform (provided that you remember that by "components transform" you mean finding the components of the same vector in the new coordinate system, and by "basis vectors transform" you mean rotating the basis vectors to point along different coordinate axes).

You should also remember that rotation of Cartesian coordinate axes is only one among many possible forms of transformation. In general, any time you choose to switch from one set of basis vectors to another, you must consider the effect of your choice of new basis vectors on the components of the vectors in your system. How the matrix that transforms the original basis vectors into the new ones relates to the matrix that converts the vector components depends on the type of component you're using to represent the vector.

If you're surprised to learn that there can be more than one type of component for a given vector, you should consider a coordinate system in which the axes are not perpendicular to one another. You can learn about such "non-orthogonal" coordinate systems in the next section.

4.4 Non-orthogonal coordinate systems

In Cartesian coordinate systems, there's no chance for ambiguity when you consider the process of "projection" of a vector onto a coordinate axis. Using the light source and shadow approach described in Chapter 1, you simply imagine a source of light shining on the vector and the shadow produced by that vector on one of the coordinate axes, as in Figure 1.6. In two-dimensional Cartesian coordinates, the direction of the light may be specified in one of two equivalent ways: parallel to one of the axes (actually antiparallel since the light

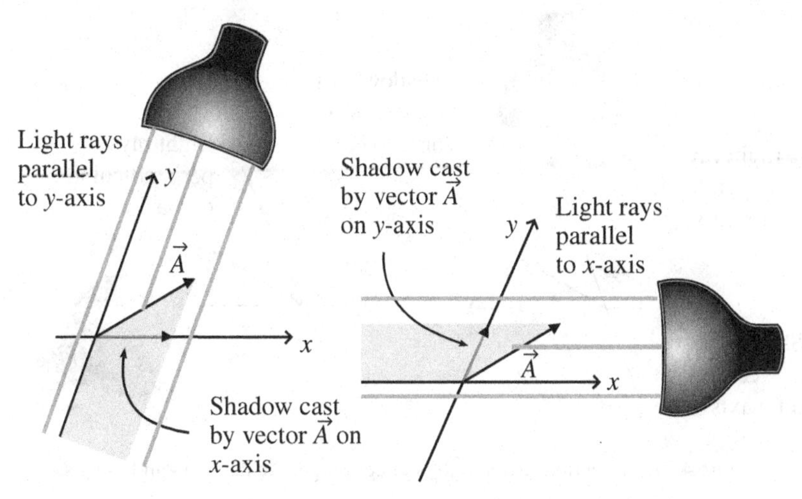

Figure 4.10 Projections using light sources parallel to x- and y-axes.

shines back toward the origin), or perpendicular to the other axis. For example, in Figure 1.6(a), you're saying exactly the same thing if you describe the light as shining "antiparallel to the y-axis" or "perpendicular to the x-axis."

Now imagine a two-dimensional coordinate system in which the x- and y-axes are not perpendicular to one another.[4] In such cases, the process of projecting a vector onto one of the coordinate axes takes on an additional complication. Should the light sources shine (anti-) parallel to the coordinate axes, as in Figure 4.10, or perpendicular to the axes, as in Figure 4.11?

In each case, a "projection" of the vector is formed onto one of the coordinate axes, but those projections may have quite different lengths, as you can see by comparing the lengths of the "shadows" cast in Figure 4.10 to those in Figure 4.11.

You may certainly be forgiven for thinking "So what?" when confronted with these differing projections. Does it really matter that there are two ways to project a vector onto an axis in non-orthogonal coordinate systems?

One indication that the type of projection does matter comes about if you attempt to use vector addition to form vector $\vec{A}$ from the projection components using the rules of vector addition. As you can see in Figure 4.12, that process works perfectly if you use the parallel-projection components but fails miserably when you attempt to use the perpendicular-projection components.

[4] This is not just an academic exercise; non-orthogonal coordinate axes turn up quite naturally in problems in relativity, fluid dynamics, and other areas.

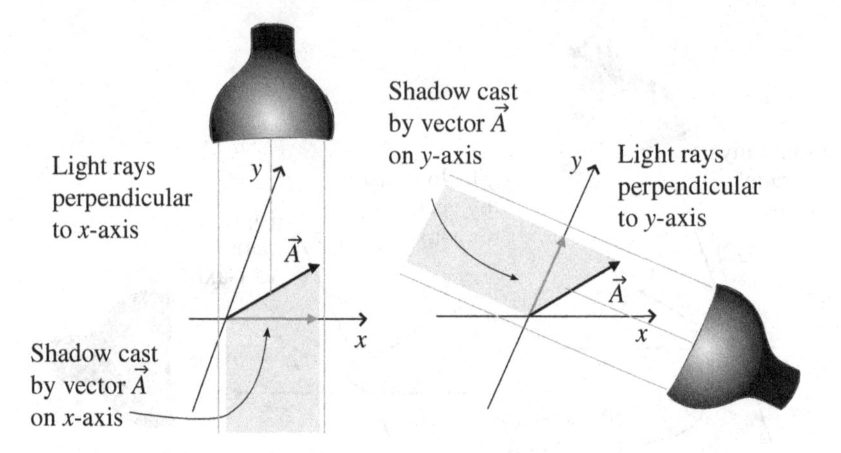

Figure 4.11 Projections using light sources perpendicular to x- and y-axes.

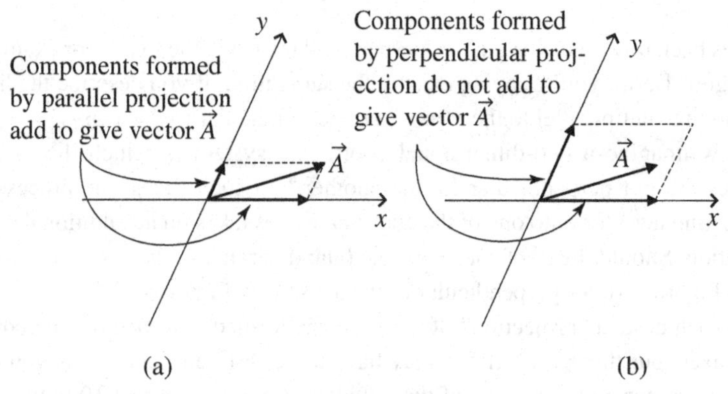

Figure 4.12 Vector addition of components formed by parallel and perpendicular projection.

This may cause you to wonder why the perpendicular-projection components are called "components" at all.

Another way to appreciate the significance of the difference between parallel and perpendicular projections is to consider how the components formed by these two types of projection transform between coordinate systems. As you'll see later in this chapter, the components formed by projections perpendicular to the coordinate axes transform between coordinate systems using the direct transformation matrix that is also used to form the new basis vectors in the new coordinate system, while the components formed by projections parallel to the coordinate axes transform between coordinate systems using the inverse transformation matrix. This behavior has caused the perpendicular-projection components to traditionally be called the "covariant" components of the vector, while the parallel-projection components are called the "contravariant" components of the vector. Of course, for orthogonal coordinate systems, the direction parallel to one of the coordinate axes is exactly the same as the direction perpendicular to other axes, so in that case the covariant and contravariant components of a vector are identical, and no distinction is needed.

To learn why the covariant values are called "components," and, much more importantly, to understand why covariant and contravariant components are meaningful quantities and how they may be used to write physical laws that do not depend on the reference frame of the observer, you should first understand the concept of dual basis vectors. You can read about such basis vectors in the next section.

4.5 Dual basis vectors

For non-orthogonal coordinate systems, it's clear from geometric considerations such as those illustrated in Figure 4.12 that the perpendicular projections of a vector onto the coordinate axes do not form "components" in the way that parallel projections do; the perpendicular projections simply don't add up as vectors to give the original vector. But to truly understand the process of "adding up" components as vectors, you have to think about the role of the basis vectors in that addition. To see how that works for parallel projections, take a look at the basis vectors $\vec{e}_1$ and $\vec{e}_2$ pointing along the (non-orthogonal) coordinate axes in Figure 4.13 and the projections of vector $\vec{A}$ onto those directions. In this case, vector $\vec{A}$ may be written as

$$\vec{A} = A^x \vec{e}_1 + A^y \vec{e}_2, \tag{4.20}$$

where A^x and A^y represent the parallel-projection (contravariant) components of $\vec{A}$.[5]

The same approach doesn't work for the perpendicular-projection (covariant) components A_x and A_y, as you can tell by looking at the lengths of the projections in Figure 4.12(b); it's clear that those two "components" multiplied by the basis vectors $\vec{e}_1$ and $\vec{e}_2$ do not add up to give $\vec{A}$. So it's reasonable to wonder if there are alternative basis vectors that would allow the

[5] The use of superscripts for the "x" and "y" in the contravariant components A^x and A^y is deliberate and is the standard notation for distinguishing these contravariant components from the covariant components A_x and A_y.

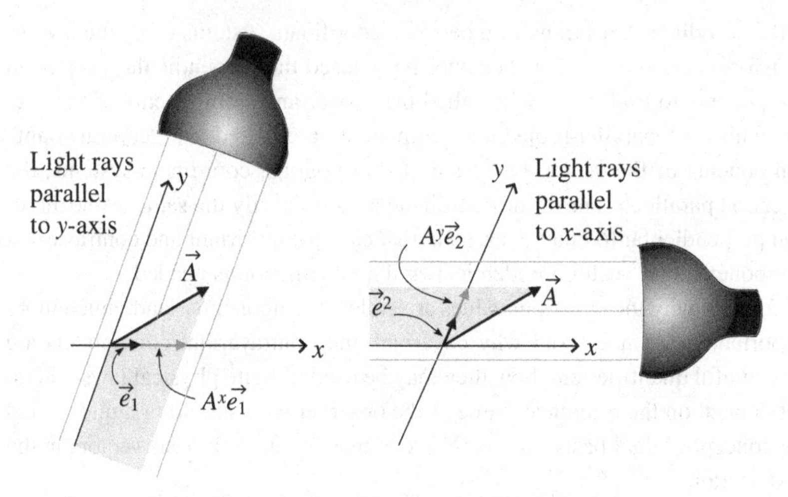

Figure 4.13 Parallel-projection components and basis vectors.

perpendicular-projection components to form a vector in a manner analogous to Eq. 4.20. Happily, there are, and those alternative basis vectors are called "reciprocal" or "dual" basis vectors.

Dual basis vectors have two defining characteristics. The first is that each one must be perpendicular to all original basis vectors with different indices. So if you call the dual basis vectors $\vec{e}^{\,1}$ and $\vec{e}^{\,2}$ to distinguish them from the original basis vectors $\vec{e}_1$ and $\vec{e}_2$, you can be sure that $\vec{e}^{\,1}$ is perpendicular to $\vec{e}_2$ (and thus perpendicular to the y-axis in this case). Likewise, $\vec{e}^{\,2}$ must be perpendicular to $\vec{e}_1$ (and thus perpendicular to the x-axis in this case). The directions of the dual basis vectors $\vec{e}^{\,1}$ and $\vec{e}^{\,2}$ are shown in Figure 4.14.

The second defining characteristic for dual basis vectors is that the dot product between each dual basis vector and the original basis vector with the same index must equal one (so $\vec{e}^{\,1} \circ \vec{e}_1 = 1$ and $\vec{e}^{\,2} \circ \vec{e}_2 = 1$). This means that you can find the lengths of the dual basis vectors as long as you know the lengths of the original basis vectors and the angle between each dual basis vector and the corresponding original basis vector.[6] So to find the length of $\vec{e}^{\,1}$, you simply have to multiply the length of the original basis vector $\vec{e}_1$ by the cosine of the angle between $\vec{e}^{\,1}$ and $\vec{e}_1$ and then take the inverse of the result. Likewise, to find the length of $\vec{e}^{\,2}$, multiply the length of the original basis vector $\vec{e}_2$ by the cosine of the angle between $\vec{e}^{\,2}$ and $\vec{e}_2$ and take the inverse of that result. Thus:

$$|\vec{e}^{\,1}| = \frac{1}{|\vec{e}_1|\cos(\theta_1)}, \tag{4.21}$$

[6] Recall from Chapter 2 that $\vec{A} \circ \vec{B} = |\vec{A}||\vec{B}|\cos\theta$, where θ is the angle between $\vec{A}$ and $\vec{B}$.

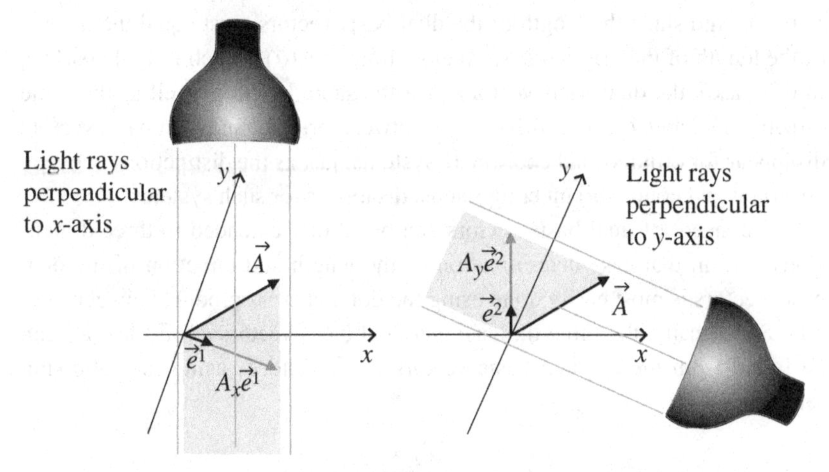

Figure 4.14 Perpendicular-projection components and dual basis vectors.

and

$$|\vec{e}^{\,2}| = \frac{1}{|\vec{e}_2|\cos(\theta_2)}, \tag{4.22}$$

where θ_1 is the angle between $\vec{e}^{\,1}$ and $\vec{e}_1$ and θ_2 is the angle between $\vec{e}^{\,2}$ and $\vec{e}_2$.

With the concept of dual basis vectors in hand, you're in a position to understand why the perpendicular-projection (covariant) components A_x and A_y may rightfully be called "components." The key is that the projections must be made onto the direction of the dual basis vectors rather than onto the directions of the original basis vectors. If you do that, then the covariant components A_x and A_y can be multiplied by the relevant basis vectors and added to give the original vector $\vec{A}$ in the same way as can be done using the parallel-projection (contravariant) components A^x and A^y. The covariant-component equivalent to Eq. 4.20 is thus

$$\vec{A} = A_x\vec{e}^{\,1} + A_y\vec{e}^{\,2}. \tag{4.23}$$

As you may have guessed, the use of superscripts to denote the dual basis vectors $\vec{e}^{\,1}$ and $\vec{e}^{\,2}$ is not accidental; when these basis vectors are transformed to a new coordinate system, the inverse transformation matrix is used, as it is for the contravariant vector components A^x and A^y.

Note that in a two-dimensional coordinate system with orthonormal basis vectors such as $\hat{\imath}$ and $\hat{\jmath}$, the dual basis vectors are identical to the original basis vectors along the coordinate axes. That's easily understood, because the direction of each of the dual basis vectors must be perpendicular to the direction of one of the original basis vectors (and hence must point along the x- and

y-axes). And since the length of the dual basis vectors must equal the inverse of the length of the original basis vectors times $\cos(\theta)$ (which is $1/[1\cos(0°)]$ in this case), the dual basis vectors have the same length as well as the same direction as $\hat{\imath}$ and $\hat{\jmath}$. So the differences between original and dual basis vectors disappear for orthonormal coordinate systems, just as the distinctions between covariant and contravariant components disappear for such systems.

The concept of dual basis vectors can be readily extended to three dimensions, and in that case determination of the length and direction of the dual basis vectors is most easily done using the dot and cross product between vectors. Specifically, the three-dimensional dual basis vectors $\vec{e}^{\,1}$, $\vec{e}^{\,2}$ and $\vec{e}^{\,3}$ can be found from the original basis vectors $\vec{e}_1$, $\vec{e}_2$, and $\vec{e}_3$ using the following relations:

$$\begin{aligned}\vec{e}^{\,1} &= \frac{\vec{e}_2 \times \vec{e}_3}{\vec{e}_1 \circ (\vec{e}_2 \times \vec{e}_3)},\\ \vec{e}^{\,2} &= \frac{\vec{e}_3 \times \vec{e}_1}{\vec{e}_1 \circ (\vec{e}_2 \times \vec{e}_3)},\\ \vec{e}^{\,3} &= \frac{\vec{e}_1 \times \vec{e}_2}{\vec{e}_1 \circ (\vec{e}_2 \times \vec{e}_3)}.\end{aligned} \tag{4.24}$$

Each denominator is the triple scalar product of the original basis vectors, which you may recall from Section 2.3 is the volume of the parallelepiped formed by those vectors.

In these equations, the cross products in the numerators ensure that the first characteristic of dual basis vectors is met (for example, that $\vec{e}^{\,1}$ is perpendicular to $\vec{e}_2$ and to $\vec{e}_3$). The triple scalar products in the denominators ensure that the second characteristic is met (for example, that $\vec{e}^{\,1} \circ \vec{e}_1 = 1$).

The computation of dual basis vectors may seem like a long trek to make simply to have an alternative way of writing vectors, but there's a great truth to be found by comparing Eqs. 4.20 and 4.23. Since these equations describe the same vector, you may combine them to write

$$\vec{A} = A^x \vec{e}_1 + A^y \vec{e}_2 = A_x \vec{e}^{\,1} + A_y \vec{e}^{\,2}, \tag{4.25}$$

which serves to emphasize an important fact. If you seek to define a quantity (such as vector $\vec{A}$) that remains invariant under a transformation of coordinates, you have a choice: you can combine superscripted (contravariant) components with subscripted (covariant) basis vectors, or you can combine subscripted (covariant) components with superscripted (contravariant) basis vectors. That should seem reasonable to you, because covariant quantities transform using a direct transformation matrix, while contravariant quantities use an inverse transformation matrix. Multiplying such quantities guarantees that the result is unaffected by the transformation.

You can see an example of how dual basis vectors and covariant and contravariant components are determined in the next section.

4.6 Finding covariant and contravariant components

Once you grasp the concept of dual basis vectors in non-orthonormal coordinate systems, finding the covariant and contravariant components of a vector is straightforward. As an example, take a look at vector $\vec{A}$ in Figure 4.15, with non-orthogonal basis vectors $\vec{e}_1$ and $\vec{e}_2$.

Finding the contravariant components A^1 and A^2 is simply a matter of parallel-projecting vector $\vec{A}$ onto the directions of the original basis vectors $\vec{e}_1$ and $\vec{e}_2$, as shown in Figure 4.16. A quick visual inspection suggests that component $A^1|\vec{e}_1|$ should be about 2/3 the length of original basis vector $\vec{e}_1$,

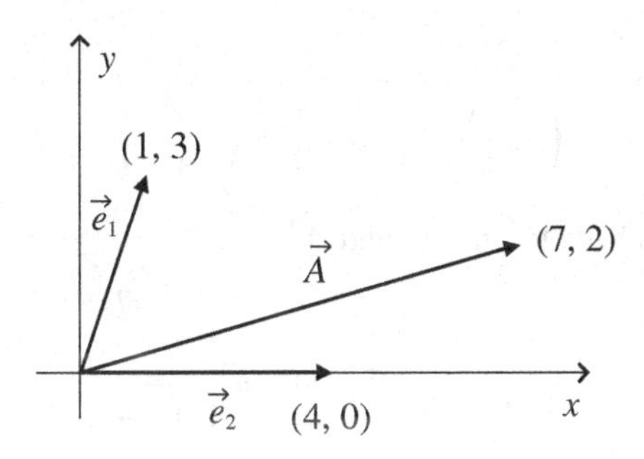

Figure 4.15 Non-orthogonal basis vectors.

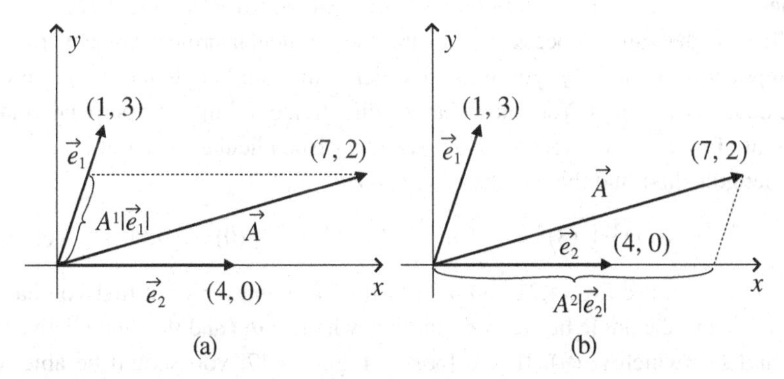

Figure 4.16 Parallel projections onto original basis vectors.

and component $A^2|\vec{e}_2|$ should be about 1.5 times the length of original basis vector $\vec{e}_2$. The values of A^1 and A^2 can be found by writing the vector equation

$$\vec{A} = A^1\vec{e}_1 + A^2\vec{e}_2, \tag{4.26}$$

which can be written as two equations for the components of $\vec{A}$:

$$A_x = A^1 e_{1,x} + A^2 e_{2,x},$$
$$A_y = A^1 e_{1,y} + A^2 e_{2,y}.$$

These two simultaneous equations may readily be solved for A^1 and A^2 using the elimination or substitution method (both of which are demonstrated in the on-line solutions to the problems at the end of this chapter). Another approach is the matrix method and Cramer's Rule (described in the matrix-algebra review on the book's website). Using this approach, you begin by substituting the known values for vector $\vec{A}$ as well as $\vec{e}_1$ and $\vec{e}_2$:

$$\begin{pmatrix} 7 \\ 2 \end{pmatrix} = A^1 \begin{pmatrix} 1 \\ 3 \end{pmatrix} + A^2 \begin{pmatrix} 4 \\ 0 \end{pmatrix}, \tag{4.27}$$

which may also be written as

$$\begin{pmatrix} 7 \\ 2 \end{pmatrix} = \begin{pmatrix} 1 & 4 \\ 3 & 0 \end{pmatrix} \begin{pmatrix} A^1 \\ A^2 \end{pmatrix}. \tag{4.28}$$

Now use Cramer's Rule to find A^1 and A^2:

$$A^1 = \frac{\begin{vmatrix} 7 & 4 \\ 2 & 0 \end{vmatrix}}{\begin{vmatrix} 1 & 4 \\ 3 & 0 \end{vmatrix}} = \frac{-8}{-12} = 0.667, \quad A^2 = \frac{\begin{vmatrix} 1 & 7 \\ 3 & 2 \end{vmatrix}}{\begin{vmatrix} 1 & 4 \\ 3 & 0 \end{vmatrix}} = \frac{-19}{-12} = 1.583. \tag{4.29}$$

These values are consistent with the visual estimates from Figure 4.16.

To use the same process to find the perpendicular-projection (covariant) components A_1 and A_2, you must first determine the length and direction of the dual basis vectors. You know that the direction of $\vec{e}^{\,1}$ must be perpendicular to that of $\vec{e}_2$, and the direction of $\vec{e}^{\,2}$ must be perpendicular to that of $\vec{e}_1$. As for the lengths, first find the lengths of $\vec{e}_1$ and $\vec{e}_2$:

$$|\vec{e}_1| = \sqrt{(1)^2 + (3)^2} = 3.16, \quad |\vec{e}_2| = \sqrt{(4)^2 + (0)^2} = 4.00. \tag{4.30}$$

Then you can use Eqs. 4.21 and 4.22 to find $|\vec{e}^{\,1}|$ and $|\vec{e}^{\,2}|$, but first you have to figure out the angle between $\vec{e}_1$ and $\vec{e}^{\,1}$ (which is θ_1) and the angle between $\vec{e}_2$ and $\vec{e}^{\,2}$ (which is θ_2). If you look at Figure 4.17, you should be able to determine that $\theta_1 = \theta_2 = \arctan(1/3) = 18.43°$, so you have

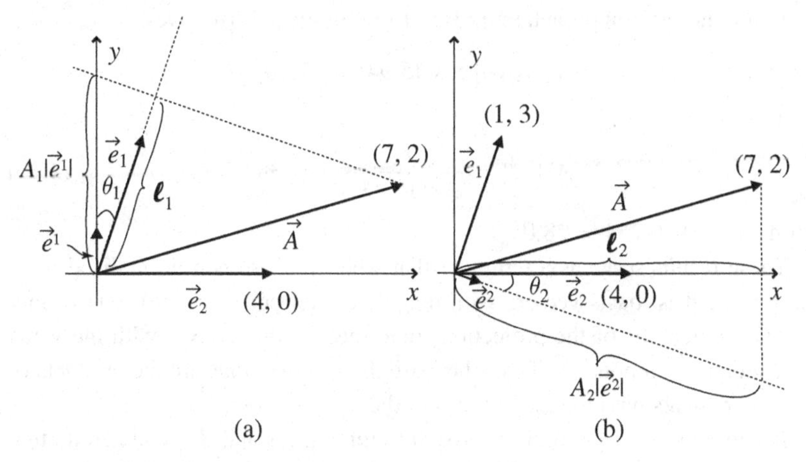

Figure 4.17 Perpendicular projections onto dual basis vectors.

$$
\begin{aligned}
|\vec{e}^{\,1}| &= \frac{1}{|\vec{e}_1|\cos(\theta_1)} = \frac{1}{3.16\cos(18.43°)} = 0.333, \\
|\vec{e}^{\,2}| &= \frac{1}{|\vec{e}_2|\cos(\theta_2)} = \frac{1}{4.00\cos(18.43°)} = 0.264.
\end{aligned}
\tag{4.31}
$$

You can see the (very short) dual basis vectors $\vec{e}^{\,1}$ and $\vec{e}^{\,2}$ in Figure 4.17. Note that $\vec{e}^{\,1}$ is perpendicular to $\vec{e}_2$ and that $\vec{e}^{\,2}$ is perpendicular to $\vec{e}_1$, and their lengths are given by Eq. 4.31.

Once you have the dual basis vectors in hand, you're in a position to find the perpendicular-projection (covariant) components A_1 and A_2. You can do this geometrically by continuing the perpendicular-projection lines beyond the direction lines of $\vec{e}_1$ and $\vec{e}_2$ and onto the direction lines of $\vec{e}^{\,1}$ and $\vec{e}^{\,2}$, as shown in Figure 4.17. The magnitude of vector $\vec{A}$ is

$$
|\vec{A}| = \sqrt{(7)^2 + (2)^2} = 7.28, \tag{4.32}
$$

and the angle between $\vec{A}$ and the x-axis is $\arctan(\frac{2}{7}) = 15.94°$. Using this value and θ_1 from above, you can determine that the angle between $\vec{A}$ and $\vec{e}_1$ is 55.62° and the angle between $\vec{A}$ and $\vec{e}_2$ is 15.94°. So the length of ℓ_1 in Figure 4.17(a) is

$$
\ell_1 = |\vec{A}|\cos(55.62°) = 4.11, \tag{4.33}
$$

and

$$
A_1|\vec{e}^{\,1}| = \frac{\ell_1}{\cos(18.43°)} = 4.33, \tag{4.34}
$$

so $A_1 = 4.33/0.333 = 13.0$.

Using the same approach to find A_2 from Figure 4.17(b) gives

$$\ell_2 = |\vec{A}| \cos(15.94°) = 7.00, \tag{4.35}$$

and

$$A_2|\vec{e}^{\,2}| = \frac{\ell_2}{\cos(18.43°)} = 7.38, \tag{4.36}$$

so $A_2 = 7.38/0.264 = 28.0$.

These results serve as a reminder that when you use non-normalized basis vectors (that is, basis vectors with magnitude not equal to one), you cannot equate the lengths of the projections onto the coordinate axes with the value of a vector's components. That's because those projections are the products of the components with the magnitudes of the basis vectors.

If you prefer the algebraic approach to finding A_1 and A_2, you can do that by proceeding as you did for A^1 and A^2, although in this case you begin with

$$\vec{A} = A_1\vec{e}^{\,1} + A_2\vec{e}^{\,2}, \tag{4.37}$$

and then substitute the known values for vector $\vec{A}$ as well as the x- and y-components of the dual basis vectors $\vec{e}^{\,1}$ and $\vec{e}^{\,2}$:

$$e^1_x = |\vec{e}^{\,1}|\cos(90°) = 0.000, \quad e^2_x = |\vec{e}^{\,2}|\cos(360° - 18.43°) = 0.250,$$
$$e^1_y = |\vec{e}^{\,1}|\sin(90°) = 0.333, \quad e^2_y = |\vec{e}^{\,2}|\sin(360° - 18.43°) = -0.083.$$

So

$$\begin{pmatrix} 7 \\ 2 \end{pmatrix} = A_1 \begin{pmatrix} 0 \\ 0.333 \end{pmatrix} + A_2 \begin{pmatrix} 0.25 \\ -0.083 \end{pmatrix}. \tag{4.38}$$

As before, this may be written as

$$\begin{pmatrix} 7 \\ 2 \end{pmatrix} = \begin{pmatrix} 0 & 0.25 \\ 0.333 & -0.083 \end{pmatrix} \begin{pmatrix} A_1 \\ A_2 \end{pmatrix}. \tag{4.39}$$

Again using Cramer's Rule to solve for A_1 and A_2 gives

$$A_1 = \frac{\begin{vmatrix} 7 & 0.25 \\ 2 & -0.083 \end{vmatrix}}{\begin{vmatrix} 0 & 0.25 \\ 0.333 & -0.083 \end{vmatrix}} = \frac{-1.081}{-0.083} = 13.0,$$

$$A_2 = \frac{\begin{vmatrix} 0 & 7 \\ 0.333 & 2 \end{vmatrix}}{\begin{vmatrix} 0 & 0.25 \\ 0.333 & -0.083 \end{vmatrix}} = \frac{-2.331}{-0.083} = 28.0, \tag{4.40}$$

as expected from the geometric approach.

A simpler approach to finding the contravariant and covariant components of a vector once you have both the original and dual basis vectors in hand is to use these relations:

$$A_1 = \vec{A} \circ \vec{e}_1 = A_x e_{1,x} + A_y e_{1,y} \quad A_2 = \vec{A} \circ \vec{e}_2 = A_x e_{2,x} + A_y e_{2,y}, \tag{4.41}$$

and

$$A^1 = \vec{A} \circ \vec{e}^{\,1} = A_x e^1_x + A_y e^1_y \quad A^2 = \vec{A} \circ \vec{e}^{\,2} = A_x e^2_x + A_y e^2_y. \tag{4.42}$$

In the current example, this approach gives the covariant components as

$$A_1 = (7, 2) \circ (1, 3) = (7)(1) + (2)(3) = 13,$$
$$A_2 = (7, 2) \circ (4, 0) = (7)(4) + (2)(0) = 28,$$

and

$$A^1 = (7, 2) \circ (0, 0.333) = (7)(0) + (2)(0.333) = 0.666,$$
$$A^2 = (7, 2) \circ (0.250, -0.083) = (7)(0.250) + (2)(-0.083) = 1.58,$$

in agreement with the geometric and matrix-algebra approaches taken above.

It's important for you to realize that what you've just found are the parallel-projection (contravariant) and perpendicular-projection (covariant) components of vector $\vec{A}$ with respect to the original basis vectors $\vec{e}_1$ and $\vec{e}_2$ and the dual basis vectors $\vec{e}^{\,1}$ and $\vec{e}^{\,2}$. So does that mean that $\vec{A}$ is a covariant vector or a contravariant vector?

The answer is neither (or both, if you prefer); it's not the vector itself that is contravariant or covariant, it's the set of components that you form through its parallel or perpendicular projections. As you read the literature on tensors, you're very likely to run into expressions such as "the contravariant vector $\vec{A}$" or "the covariant vector $\vec{B}$," and what the author generally means is that the contravariant components of vector $\vec{A}$ and the covariant components of vector $\vec{B}$ are being used for the problem (perhaps because they're simpler). But you can be sure that like all vectors, $\vec{A}$ and $\vec{B}$ both have contravariant and covariant components, and you can find them using the techniques described in this section.[7]

And if you're wondering why you might want to go through the effort of finding those components, rest assured that the payoff is worth the effort. To appreciate the value of that payoff, you'll have to begin thinking of vectors not just as arrows with a certain length and pointing in a specified direction, but rather as members of a class of objects called tensors that have very predictable

[7] In Chapter 5, you can learn to move between contravariant and covariant components using the metric tensor.

(and useful) properties under transformation of coordinates. In that view, the vectors you've been dealing with up to this point have all been tensors of rank one. Seeing them as such, and understanding what that means, will be made a great deal easier through the use of a notation called "index notation" and a convention known as the "Einstein summation convention." You can read about index notation and the summation convention in the next section.

4.7 Index notation

You've seen the first glimmerings of index notation in the earlier section of this chapter describing coordinate transformations. As you may recall, the angles between the transformed (rotated) coordinate axes and the original (non-rotated) axes of a two-dimensional coordinate system were called α_{11}, α_{12}, α_{21}, and α_{22}. These angles could just as well have been designated $\alpha_{x'x}$, $\alpha_{x'y}$, $\alpha_{y'x}$, and the like, but there are several good reasons to use the index numbers 1, 2, and 3 rather than the letters x, y, and z to refer to coordinate axes and vector components. One of those reasons is that many problems in physics and engineering involve a number of dimensions greater than 3, and although everyone agrees that "4" comes after "3," a consensus hasn't been reached on what comes after "z." Another reason is that index notation enables the great convenience of the summation convention that you can read about later in this section.

Using index notation, the coordinates of a point in three-dimensional space are written as (x_1, x_2, x_3) or (x^1, x^2, x^3) rather than (x, y, z), and the components of a vector are written as (A_1, A_2, A_3) or (A^1, A^2, A^3) rather than (A_x, A_y, A_z) or (A^x, A^y, A^z). This system is easily extended to N-dimensional space, in which the coordinates become $(x_1, x_2, \ldots, x_N)$ or $(x^1, x^2, \ldots, x^N)$ and the vector components become $(A_1, A_2, \ldots, A_N)$ or $(A^1, A^2, \ldots, A^N)$.

Applying this notation to the equation for the transformation of contravariant vector components produced by a rotation of two-dimensional axes, Eq. 4.6 becomes

$$\begin{pmatrix} A'^1 \\ A'^2 \end{pmatrix} = \begin{pmatrix} \cos(\alpha_{11}) & \cos(\alpha_{12}) \\ \cos(\alpha_{21}) & \cos(\alpha_{22}) \end{pmatrix} \begin{pmatrix} A^1 \\ A^2 \end{pmatrix}. \tag{4.43}$$

In three dimensions, this is

$$\begin{pmatrix} A'^1 \\ A'^2 \\ A'^3 \end{pmatrix} = \begin{pmatrix} \cos(\alpha_{11}) & \cos(\alpha_{12}) & \cos(\alpha_{13}) \\ \cos(\alpha_{21}) & \cos(\alpha_{22}) & \cos(\alpha_{23}) \\ \cos(\alpha_{31}) & \cos(\alpha_{32}) & \cos(\alpha_{33}) \end{pmatrix} \begin{pmatrix} A^1 \\ A^2 \\ A^3 \end{pmatrix}. \tag{4.44}$$

Designating the elements of the transformation matrix a_{11}, a_{12}, a_{13}, and so forth allows you to write Eq. 4.44 as

$$
\begin{aligned}
A^{'1} &= a_{11}A^1 + a_{12}A^2 + a_{13}A^3, \\
A^{'2} &= a_{21}A^1 + a_{22}A^2 + a_{23}A^3, \\
A^{'3} &= a_{31}A^1 + a_{32}A^2 + a_{33}A^3,
\end{aligned}
\tag{4.45}
$$

or

$$
\begin{aligned}
A^{'1} &= \sum_{j=1}^{3} a_{1j}A^j, \\
A^{'2} &= \sum_{j=1}^{3} a_{2j}A^j, \\
A^{'3} &= \sum_{j=1}^{3} a_{3j}A^j.
\end{aligned}
\tag{4.46}
$$

Allowing "i" to stand for any of the indices 1, 2, or 3 makes this:

$$
A^{'i} = \sum_{j=1}^{3} a_{ij}A^j. \qquad i = 1, 2, 3 \tag{4.47}
$$

As a final simplification, whenever an index appears twice in the same term, once as a superscript and once as a subscript (as "j" does in Eq. 4.47), you can omit the summation symbol and write simply

$$
A^{'i} = a_{ij}A^j, \tag{4.48}
$$

in which the reader knows to sum over the repeated index (j in this case). Such repeated indices are often called "dummy" indices, since any letter may be used for that index and the result will be the same.[8] It was Albert Einstein who first suggested this summation convention, which he jokingly called his "great discovery in mathematics."[9] Whatever you call it, this idea certainly has saved a lot of ink and time since Einstein proposed it in 1916.

Before moving on, you should take a careful look at Eq. 4.48 and make sure you understand that these few symbols mean exactly the same thing as the many terms in the three separate equations of Eq. 4.45. They tell you that

[8] Unlike the repeated "dummy" indices which indicate summation, i is called a "free" index and no summation is implied.

[9] Pais, A. 1983, *Subtle Is the Lord: The Science and the Life of Albert Einstein*, Oxford University Press, Oxford.

each component in the primed coordinate system is a weighted linear combination of the components in the original (unprimed) coordinate system, with the transformation matrix elements (a_{ij}) providing the weighting factors for each term.

And if you want to know the exact meaning of each of those factors in the transformation of covariant and contravariant vector components, the next section will help with that.

4.8 Quantities that transform contravariantly

With the convenience of index notation and the summation convention at your disposal, you should be ready to take the next step in the transition from thinking of vectors as quantities with magnitude and direction to understanding why vectors belong to the class of objects known as tensors. That step begins by asking the question of how a differential element of length $d\vec{s}$ transforms from one coordinate system to another.

In general, the equations relating the coordinates in one system to those in another do not involve simple linear combinations of coordinate values. For example, in transforming from spherical (r, θ, ϕ) to Cartesian (x, y, z) coordinates, it's not possible to write equations such as $x = a_{11}r + a_{12}\theta + a_{13}\phi$, because x depends on the product of r with the sine of θ and the cosine of ϕ. And y and z have similar non-linear relationships to the spherical coordinates.

If, however, you ask how the differentials of x, y, and z (that is, dx, dy, and dz) depend on the differentials of r, θ, and ϕ (that is, dr, $d\theta$, and $d\phi$), you'll find that on this infinitesimally small scale, dx does depend linearly on dr, $d\theta$, and $d\phi$ (as do dy and dz). So you are able to write

$$dx = a_{11}dr + a_{12}d\theta + a_{13}d\phi, \tag{4.49}$$

and likewise for dy and dz.

For any two coordinate systems in which a linear relationship exists between differential length elements, writing the equations which transform between the systems is straightforward. If you call the differentials of one coordinate system dx, dy, and dz and the other coordinate system dx', dy', and dz', the transformation equations from the unprimed to the primed systems come directly from the rules of partial differentiation, as shown in the left column below:

$$
\begin{aligned}
dx' &= \frac{\partial x'}{\partial x}dx + \frac{\partial x'}{\partial y}dy + \frac{\partial x'}{\partial z}dz \Rightarrow dx'^1 = \frac{\partial x'^1}{\partial x^1}dx^1 + \frac{\partial x'^1}{\partial x^2}dx^2 + \frac{\partial x'^1}{\partial x^3}dx^3, \\
dy' &= \frac{\partial y'}{\partial x}dx + \frac{\partial y'}{\partial y}dy + \frac{\partial y'}{\partial z}dz \Rightarrow dx'^2 = \frac{\partial x'^2}{\partial x^1}dx^1 + \frac{\partial x'^2}{\partial x^2}dx^2 + \frac{\partial x'^2}{\partial x^3}dx^3, \\
dz' &= \frac{\partial z'}{\partial x}dx + \frac{\partial z'}{\partial y}dy + \frac{\partial z'}{\partial z}dz \Rightarrow dx'^3 = \frac{\partial x'^3}{\partial x^1}dx^1 + \frac{\partial x'^3}{\partial x^2}dx^2 + \frac{\partial x'^3}{\partial x^3}dx^3.
\end{aligned}
\tag{4.50}
$$

Using the index-notation approach of substituting x^1, x^2, and x^3 for x, y, and z results in the column shown on the right.[10] Putting this into matrix notation gives

$$
\begin{pmatrix} dx'^1 \\ dx'^2 \\ dx'^3 \end{pmatrix} = \begin{pmatrix} \frac{\partial x'^1}{\partial x^1} & \frac{\partial x'^1}{\partial x^2} & \frac{\partial x'^1}{\partial x^3} \\ \frac{\partial x'^2}{\partial x^1} & \frac{\partial x'^2}{\partial x^2} & \frac{\partial x'^2}{\partial x^3} \\ \frac{\partial x'^3}{\partial x^1} & \frac{\partial x'^3}{\partial x^2} & \frac{\partial x'^3}{\partial x^3} \end{pmatrix} \begin{pmatrix} dx^1 \\ dx^2 \\ dx^3 \end{pmatrix}, \tag{4.51}
$$

or, using individual equations with summation symbols

$$
dx'^1 = \sum_{j=1}^{3} \frac{\partial x'^1}{\partial x^j}dx^j, \quad dx'^2 = \sum_{j=1}^{3} \frac{\partial x'^2}{\partial x^j}dx^j, \quad dx'^3 = \sum_{j=1}^{3} \frac{\partial x'^3}{\partial x^j}dx^j.
$$

If you now allow the letter i to represent each of the numerical values of the index (1, 2, and 3), this can be written as

$$
dx'^i = \sum_{j=1}^{3} \frac{\partial x'^i}{\partial x^j}dx^j. \tag{4.52}
$$

Since the j index is repeated, a final simplification results from the Einstein summation convention, allowing you to write

$$
dx'^i = \frac{\partial x'^i}{\partial x^j}dx^j. \tag{4.53}
$$

So index notation has allowed the expression in Eq. 4.50, consisting of three equations with three terms in each, to be written as this single equation. More importantly, the form of this equation will help you understand why differential length elements (dx^i) are considered to be contravariant quantities.

[10] Superscripts are used for the indices because differential length elements transform as contravariant quantities, as described later in this section.

To gain that understanding, it's useful to recall Eq. 4.48 from the previous section:

$$A'^i = a_{ij} A^j,$$

which tells you that the components of a vector in the primed (transformed) coordinate system are the weighted linear combination of the components of that same vector in the unprimed (original) coordinate system. And the weighting factors a_{ij} are the elements of the transformation matrix.

Now compare Eq. 4.53 to Eq. 4.48. On the left side of both equations, a primed quantity (dx'^i or A'^i) with free index i appears. On the right side, both equations contain the product of a factor with free index i and dummy index j ($\frac{\partial x'^i}{\partial x^j}$ or a_{ij}) with the left-side quantity unprimed and with dummy index j (dx^j or A^j). And you know that the factor a_{ij} in Eq. 4.48 represents the elements of a transformation matrix for contravariant vector components between the unprimed and the primed coordinate systems. So it seems reasonable to conclude that the $\frac{\partial x'^i}{\partial x^j}$ terms in Eq. 4.53 can be seen as the elements of the transformation matrix for the differential length elements.

So instead of looking at Eq. 4.53 as simply the index-notation version of the chain rule, you should see it as a transformation equation that takes differential length elements from the unprimed to the primed coordinate system (just as Eq. 4.48 does for the contravariant components of vector $\vec{A}$).

And here's the important insight: the $\frac{\partial x'^i}{\partial x^j}$ terms are not only the elements of a transformation matrix from the unprimed to the primed coordinate system, they're also the components of the basis vectors tangent to the original (unprimed) coordinate axes, expressed in the new (primed) coordinate system.[11]

Furthermore, you know that basis vectors tangent to the original coordinate axes are the covariant basis vectors described earlier. And since contravariant vector components combine with covariant basis vectors to produce invariant quantities, differential length elements must transform as contravariant vector components. This is the reason that the indices are written as superscripts in Eqs. 4.51 through 4.53; the differential length element is the "prototype" of contravariant vector components.

Using index notation and representing the components of the basis vectors as $\frac{\partial x'^i}{\partial x^j}$, you should now understand why the transformation equation for contravariant components of vector $\vec{A}$ is often written as

[11] If you're wondering how partial derivatives can represent basis vectors, you should review Section 2.6 of Chapter 2.

$$A'^i = \frac{\partial x'^i}{\partial x^j} A^j. \tag{4.54}$$

Many authors present this as the definition of contravariant components.

To see how this notation works in practice, consider the transformation from polar (r, θ) to two-dimensional Cartesian (x, y) coordinates. In this case, $x'^1 = x$, $x'^2 = y$, $x^1 = r$, and $x^2 = \theta$, and you know that $x = r\cos(\theta)$ and $y = r\sin(\theta)$. So what are the weighting factors (that is, the elements of the transformation matrix) in this case? Taking the appropriate derivatives, you find that

$$\frac{\partial x'^1}{\partial x^1} = \frac{\partial x}{\partial r} = \cos(\theta), \qquad \frac{\partial x'^2}{\partial x^1} = \frac{\partial y}{\partial r} = \sin(\theta), \tag{4.55}$$

$$\frac{\partial x'^1}{\partial x^2} = \frac{\partial x}{\partial \theta} = -r\sin(\theta), \qquad \frac{\partial x'^2}{\partial x^2} = \frac{\partial y}{\partial \theta} = r\cos(\theta). \tag{4.56}$$

Are these really the components of the tangent vectors to the original (r, θ) coordinate axes (that is, are they pointing along those axes)? You can see that they are by writing these terms as components in the primed coordinate system (Cartesian in this case):

$$\vec{e}_1 = \frac{\partial x'^1}{\partial x^1}\hat{\imath} + \frac{\partial x'^2}{\partial x^1}\hat{\jmath} = \cos(\theta)\hat{\imath} + \sin(\theta)\hat{\jmath}, \tag{4.57}$$

$$\vec{e}_2 = \frac{\partial x'^1}{\partial x^2}\hat{\imath} + \frac{\partial x'^2}{\partial x^2}\hat{\jmath} = -r\sin(\theta)\hat{\imath} + r\cos(\theta)\hat{\jmath}. \tag{4.58}$$

The first of these expressions is a vector pointing radially outward (along the $\hat{r}$-direction in polar coordinates) and the second is a vector pointing perpendicular to the radial direction (along the $\hat{\theta}$-direction).[12] This demonstrates that the partial derivatives in Eq. 4.53 do indeed represent components of the original (unprimed) covariant basis vectors expressed in the new (primed) coordinate system.

4.9 Quantities that transform covariantly

If the differential length element of the previous section serves as the "prototype" for quantities that transform as contravariant vector components, you may be wondering if there's a similar "prototype" for covariant quantities. You can answer that question by considering a quantity such as the change in temperature with distance (degrees per meter) over some region, which you may recognize from Chapter 2 as the gradient of that quantity. Unlike

[12] These basis vectors can be understood in terms of the non-Cartesian unit vectors discussed in Section 1.5 of Chapter 1.

the differential length element, which has dimensions directly related to the coordinate dimensions, quantities such as the gradient have dimensions that include the *inverse* of the coordinate dimensions (per unit length rather than length in the case of spatial coordinates). This dimensional consideration suggests that the gradient may be a good candidate for the prototype of quantities that transform as covariant vector components. And index notation makes this easy to see.

Imagine a scalar quantity such as temperature or density whose value at various positions is given by the function $f(x, y, z)$; the rate of change of that quantity is $\frac{\partial f}{\partial x}$ in the x-direction, $\frac{\partial f}{\partial y}$ in the y-direction, and $\frac{\partial f}{\partial z}$ in the z-direction. It's reasonable to ask how these rates of change vary if the coordinate system is changed. To answer that question, you can proceed as we did for the differential length element, using the chain rule for partial derivatives and then employing index notation as follows:

$$\frac{\partial f}{\partial x'} = \frac{\partial f}{\partial x}\frac{\partial x}{\partial x'} + \frac{\partial f}{\partial y}\frac{\partial y}{\partial x'} + \frac{\partial f}{\partial z}\frac{\partial z}{\partial x'}$$

$$\Rightarrow \frac{\partial f}{\partial x'^1} = \frac{\partial f}{\partial x^1}\frac{\partial x^1}{\partial x'^1} + \frac{\partial f}{\partial x^2}\frac{\partial x^2}{\partial x'^1} + \frac{\partial f}{\partial x^3}\frac{\partial x^3}{\partial x'^1},$$

$$\frac{\partial f}{\partial y'} = \frac{\partial f}{\partial x}\frac{\partial x}{\partial y'} + \frac{\partial f}{\partial y}\frac{\partial y}{\partial y'} + \frac{\partial f}{\partial z}\frac{\partial z}{\partial y'}$$

$$\Rightarrow \frac{\partial f}{\partial x'^2} = \frac{\partial f}{\partial x^1}\frac{\partial x^1}{\partial x'^2} + \frac{\partial f}{\partial x^2}\frac{\partial x^2}{\partial x'^2} + \frac{\partial f}{\partial x^3}\frac{\partial x^3}{\partial x'^2},$$

$$\frac{\partial f}{\partial z'} = \frac{\partial f}{\partial x}\frac{\partial x}{\partial z'} + \frac{\partial f}{\partial y}\frac{\partial y}{\partial z'} + \frac{\partial f}{\partial z}\frac{\partial z}{\partial z'}$$

$$\Rightarrow \frac{\partial f}{\partial x'^3} = \frac{\partial f}{\partial x^1}\frac{\partial x^1}{\partial x'^3} + \frac{\partial f}{\partial x^2}\frac{\partial x^2}{\partial x'^3} + \frac{\partial f}{\partial x^3}\frac{\partial x^3}{\partial x'^3}.$$

As before, you can write this as a matrix equation

$$\begin{pmatrix} \frac{\partial f}{\partial x'^1} \\ \frac{\partial f}{\partial x'^2} \\ \frac{\partial f}{\partial x'^3} \end{pmatrix} = \begin{pmatrix} \frac{\partial x^1}{\partial x'^1} & \frac{\partial x^2}{\partial x'^1} & \frac{\partial x^3}{\partial x'^1} \\ \frac{\partial x^1}{\partial x'^2} & \frac{\partial x^2}{\partial x'^2} & \frac{\partial x^3}{\partial x'^2} \\ \frac{\partial x^1}{\partial x'^3} & \frac{\partial x^2}{\partial x'^3} & \frac{\partial x^3}{\partial x'^3} \end{pmatrix} \begin{pmatrix} \frac{\partial f}{\partial x^1} \\ \frac{\partial f}{\partial x^2} \\ \frac{\partial f}{\partial x^3} \end{pmatrix}, \tag{4.59}$$

or as individual equations using the summation symbol:

$$\frac{\partial f}{\partial x'^1} = \sum_{j=1}^{3} \frac{\partial x^j}{\partial x'^1}\frac{\partial f}{\partial x^j}, \quad \frac{\partial f}{\partial x'^2} = \sum_{j=1}^{3} \frac{\partial x^j}{\partial x'^2}\frac{\partial f}{\partial x^j}, \quad \frac{\partial f}{\partial x'^3} = \sum_{j=1}^{3} \frac{\partial x^j}{\partial x'^3}\frac{\partial f}{\partial x^j}.$$

Once again employing i as the free index gives

$$\frac{\partial f}{\partial x'^i} = \sum_{j=1}^{3} \frac{\partial x^j}{\partial x'^i}\frac{\partial f}{\partial x^j}, \tag{4.60}$$

and the Einstein summation convention simplifies this to

$$\frac{\partial f}{\partial x'^i} = \frac{\partial x^j}{\partial x'^i}\frac{\partial f}{\partial x^j}. \tag{4.61}$$

Comparing this to the equivalent expression for the differential length element (Eq. 4.53) suggests that once again the vector components in the primed coordinate system are the weighted linear combination of the components in the original coordinate system. But in this case the elements of the transformation matrix ($\frac{\partial x^j}{\partial x'^i}$) are the *inverse* of those in the transformation of the differential length elements (which are $\frac{\partial x'^i}{\partial x^j}$). And just as in that case the $\frac{\partial x'^i}{\partial x^j}$ terms represent the components of vectors that point along the original coordinate axes, in this case the $\frac{\partial x^j}{\partial x'^i}$ terms represent the components of vectors that are perpendicular to the original coordinate surfaces. Hence in this case the weighting factors are the components of the (contravariant) dual basis vectors, which means that the components of the gradient vector transform as covariant components. Of course, for orthonormal coordinate systems the lengths and directions of the original and dual basis vectors are exactly the same, and there is no difference between the covariant and contravariant vector components. In non-orthonormal coordinate systems, this distinction is critically important.

Again using index notation and representing the dual basis vectors as $\frac{\partial x^j}{\partial x'^i}$, you probably won't find it surprising that many authors define the covariant components of vector $\vec{A}$ as components that transform according to the equation

$$A'_i = \frac{\partial x^j}{\partial x'^i} A_i. \tag{4.62}$$

At this point you should be convinced that vectors are more than just little arrows with magnitude and direction; they're quantities that transform in certain ways between coordinate systems. Specifically, every vector has both contravariant and covariant components that transform in predictable ways.

The contravariant components vary in the opposite manner to the basis vectors pointing along the original coordinate axes, and the covariant components vary in the same manner as those basis vectors. Most importantly, by combining the vector's contravariant components with the original basis vectors, or by combining the vector's covariant components with the dual basis vectors, the resulting quantity (the vector itself) remains invariant under all coordinate transformations. It is this characteristic that qualifies vectors to join the ranks of tensors.

Understanding the distinction between contravariant and covariant vector components is extremely helpful in understanding tensors, because vectors *are* tensors. Specifically, since all the components of a vector can be delineated using only a single index, vectors are tensors of rank one. Under this definition, scalars are tensors of rank zero, since scalars are single numbers and require no index at all. And of what use are tensors of rank two and higher? You'll encounter those in Chapter 5.

4.10 Chapter 4 problems

4.1 Write the inverse transformation matrix for a 70° rotation of the 2-D Cartesian coordinate axes and the indirect transformation matrix for the rotation of a vector through an angle of 70° degrees. Show that the product of these two transformation matrices is the identity matrix.

4.2 Use the inverse transformation matrix from Problem 4.1 to find the components of vector $\vec{A} = 2\hat{\imath} + 5.5\hat{\jmath}$ in the rotated coordinate system.

4.3 Use the direct transformation matrix from Problem 4.1 to rotate the original coordinate basis vectors $\hat{\imath}$ and $\hat{\jmath}$ by 70°, so they point along the rotated axes.

4.4 Use a direct transformation matrix to rotate vector $\vec{A}$ from Problem 4.2 through an angle of $-70°$, and compare the x- and y-components of the rotated vector (in the original coordinate system) to the x'- and y'-components of the unrotated vector in the rotated coordinate system.

4.5 Use the dot product of the original vector $\vec{A}$ with the rotated basis vectors ($\vec{A} \circ \hat{\imath}'$ and $\vec{A} \circ \hat{\jmath}'$) to find the components of $\vec{A}$ in the rotated coordinate system.

4.6 For vector $\vec{A} = -5\hat{\imath} + 6\hat{\jmath}$ and basis vectors $\vec{e}_1 = \hat{\imath} + 2\hat{\jmath}$ and $\vec{e}_2 = -2\hat{\imath} - \hat{\jmath}$, find the contravariant components $\vec{A}^1$ and $\vec{A}^2$.

4.7 Find the dual basis vectors $\vec{e}^{\,1}$ and $\vec{e}^{\,2}$ for the basis vectors $\vec{e}_1$ and $\vec{e}_2$ of Problem 4.6.

4.8 Find the covariant components $\vec{A}_1$ and $\vec{A}_2$ for vector $\vec{A}$ of Problem 4.6.

4.9 Use the subsitution method and the elimination method to solve the two simultaneous equations that result from vector Eq. 4.26.

4.10 Show that the elements of the Cartesian-to-polar transformation matrix are the components of the basis vectors tangent to the original (Cartesian) coordinate axes.

5

Higher-rank tensors

The previous chapter contains several ideas that are important to a full understanding of tensors. The first is that any vector may be represented by components that transform between coordinate systems in one of two ways. "Covariant" components transform in the same manner as the original basis vectors pointing along the coordinate axes, and "contravariant" components transform in the inverse manner of those basis vectors.[1] The second main idea is that coordinate basis vectors are tangent to the coordinate axes, and that there also exist reciprocal or dual basis vectors that are perpendicular to the coordinate axes; these dual basis vectors transform inversely to the coordinate basis vectors. The third idea is that combining contravariant components with original basis vectors and combining covariant components with dual basis vectors produces a result that is invariant under coordinate transformation. That result is the vector itself, and the vector is the same no matter which coordinate system you use for its components.

This chapter extends the concepts of covariance and contravariance beyond vectors and makes it clear that scalars and vectors are members of the class of objects called "tensors."

5.1 Definitions (advanced)

In the basic definitions of Chapter 1, scalars, vectors, and tensors were defined by the number of directions involved: zero for scalars, one for vectors, and more than one for tensors.[2] Now that you've seen the concepts of components, basis vectors, and the transformation properties of each, you're in a position

[1] The prototype of a vector expressed in contravariant components is the displacement vector, and the prototype of a vector expressed in covariant components is the gradient vector.
[2] Note that specifying one direction in 3-dimensional space requires two angles.

to understand the more-advanced definitions of scalars, vectors, and tensors. Specifically:

A scalar is a single value with no directional indicator that represents a quantity that does not vary as the coordinate system is changed.

So for a scalar with value ϕ in one coordinate system and value ϕ' in another coordinate system, you can be certain that the quantity represented by ϕ (combined with the relevant unit) and ϕ' (combined with its unit) is the same no matter which system you use to represent it. Thus 1 inch and 2.54 centimeters represent the same quantity of length.

A vector is an array of three values (in 3-D space) called "vector components" that combine with directional indicators ("basis vectors") to form a quantity that does not vary as the coordinate system is changed.

So vector $\vec{A}$ represents the same entity whether it is expressed using contravariant components A^i or covariant components A_i:

$$\vec{A} = A^i \vec{e}_i = A_i \vec{e}^{\,i},$$

where $\vec{e}_i$ represents a covariant basis vector and $\vec{e}^{\,i}$ represents a contravariant basis vector.

In transforming between coordinate systems, a vector with contravariant components A^j in the original (unprimed) coordinate system and contravariant components A'^i in the new (primed) coordinate system transforms as

$$A'^i = \frac{\partial x^{i'}}{\partial x^j} A^j,$$

where the $\frac{\partial x^{i'}}{\partial x^j}$ terms represent the components in the new coordinate system of the basis vectors tangent to the original axes.

Likewise, for a vector with covariant components A_j in the original (unprimed) coordinate system and covariant components A'_i in the new (primed) coordinate system, the transformation equation is

$$A'_i = \frac{\partial x^j}{\partial x^{i'}} A_j,$$

where the $\frac{\partial x^j}{\partial x^{i'}}$ terms represent the components in the new coordinate system of the (dual) basis vectors perpendicular to the original axes.

A tensor of rank n is an array of 3^n values (in 3-D space) called "tensor components" that combine with multiple directional indicators (basis vectors) to form a quantity that does not vary as the coordinate system is changed.

From this definition, you can see that a second-rank tensor has $3^2 = 9$ components in three-dimensional space. Note that a tensor of rank 0 is a scalar and a tensor of rank 1 is a vector.

There is no standard notation for tensors; you may see a tensor represented with double overhead arrows (such as $\vec{\vec{T}}$) or with a tilde or two-directional arrow above or below (such as $\tilde{T}$, $\overleftrightarrow{T}$ or $\underleftrightarrow{T}$). Many authors don't bother with arrows or tildes and represent tensors simply by writing the letter signifying the tensor with "placeholder" indices to indicate the contravariant and covariant rank of the tensor (such as T^{ij} or T^a_b).

5.2 Covariant, contravariant, and mixed tensors

You should by this point understand that the expression

$$A'^i = \frac{\partial x^{i'}}{\partial x^j} A^j \tag{5.1}$$

presents the contravariant components of vector $\vec{A}$ in the transformed (primed) coordinate system (A'^i) as a weighted sum of the components of $\vec{A}$ in the original (unprimed) coordinate system (A^j). The weighting factors ($\frac{\partial x^{i'}}{\partial x^j}$) are simply the elements of the transformation matrix from the unprimed to the primed coordinate systems, and those elements represent the components of the basis vectors tangent to the original coordinate axes. With that understanding, a tensor expression such as

$$A'^{ij} = \frac{\partial x^{i'}}{\partial x^k} \frac{\partial x^{j'}}{\partial x^l} A^{kl} \tag{5.2}$$

should have some recognizable elements. As you can probably surmise, in this expression A'^{ij} are the contravariant tensor components in the new coordinate system, A^{kl} are the contravariant tensor components in the original coordinate system, and $\frac{\partial x^{i'}}{\partial x^k}$ as well as $\frac{\partial x^{j'}}{\partial x^l}$ are elements of the transformation matrix between the original and new coordinate systems. And just as in Eq. 5.1, the elements of the direct transformation matrix also represent the basis vectors tangent to the original coordinate axes. But in the vector expression Eq. 5.1 each component pertains to a single basis vector, whereas the components in the tensor expression Eq. 5.2 pertain to two basis vectors. This should seem reasonable to you, since the basic definitions in Chapter 1 state that vectors involve a single direction while higher-rank tensors involve two or more directions.

The vector Eq. 5.1 involves contravariant components (as indicated by the use of superscripted indices in A'^i and A^j), but you know that an equivalent expression exists for the covariant components:

$$A'_i = \frac{\partial x^j}{\partial x^{i'}} A_j. \tag{5.3}$$

In this equation, the covariant components of vector $\vec{A}$ in the transformed (primed) coordinate system (A'_i) are expressed as a weighted sum of the covariant components of $\vec{A}$ in the original (unprimed) coordinate system (A_j). In this case, the weighting factors ($\frac{\partial x^j}{\partial x^{i'}}$) are the elements of the inverse transformation matrix from the unprimed to the primed coordinate systems, and those elements represent the dual basis vectors perpendicular to the original coordinate axes.

Extending this to a second-rank tensor gives a transformation equation such as this:

$$A'_{ij} = \frac{\partial x^k}{\partial x^{i'}} \frac{\partial x^l}{\partial x^{j'}} A_{kl}. \tag{5.4}$$

In this expression, A'_{ij} are the covariant tensor components in the new coordinate system, A_{kl} are the covariant tensor components in the original coordinate system, and $\frac{\partial x^k}{\partial x^{i'}}$ as well as $\frac{\partial x^l}{\partial x^{j'}}$ are elements of the transformation matrix between the original and new coordinate systems. And much as in Eq. 5.3, the elements of the transformation matrix represent the dual basis vectors perpendicular to the original coordinate axes.

As you may have anticipated, another possibility exists for second-rank tensors:

$$A'^i_j = \frac{\partial x^{i'}}{\partial x^k} \frac{\partial x^l}{\partial x^{j'}} A^k_l, \tag{5.5}$$

in which the tensor $\vec{\vec{A}}$ is represented by one contravariant and one covariant index; each uses the transformation matrix appropriate for its type.

5.3 Tensor addition and subtraction

As you may recall from Section 1.4, two or more vectors can be added simply by adding their corresponding components. Hence a single vector equation such as

$$\vec{C} = \vec{A} + \vec{B}, \tag{5.6}$$

actually consists of three equations (in three-dimensional space), since each component of the resultant vector $\vec{C}$ must be the sum of the corresponding components of vectors $\vec{A}$ and $\vec{B}$:

$$\begin{aligned} C_x &= A_x + B_x, \\ C_y &= A_y + B_y, \\ C_z &= A_z + B_z. \end{aligned} \tag{5.7}$$

Higher-order tensors can be added using the same process, provided that the tensors to be added have the same structure (that is, they are the same order and have the same number of covariant indices and the same number of contravariant indices). The result of tensor addition is also a tensor, and the resultant tensor has the same structure as each of the tensors that are added:

$$\begin{aligned} C_{ij} &= A_{ij} + B_{ij}, \\ C^{ij} &= A^{ij} + B^{ij}, \\ C^i_j &= A^i_j + B^i_j. \end{aligned} \tag{5.8}$$

Note that each of these expressions represents more than one equation; the exact number depends on the number of values that each index may take on. Note also that you can add tensors with any number of covariant and contravariant indices, as long as the tensors being added have the same number of each type of index.

To see that the result of adding two tensors fits the definition of a tensor, consider how the tensor components A^i_j and B^i_j transform to another coordinate system:

$$\begin{aligned} A'^k_l &= \frac{\partial x'^k}{\partial x^i}\frac{\partial x^j}{\partial x^{l'}} A^i_j, \\ B'^k_l &= \frac{\partial x'^k}{\partial x^i}\frac{\partial x^j}{\partial x^{l'}} B^i_j. \end{aligned} \tag{5.9}$$

Hence

$$\begin{aligned} A'^k_l + B'^k_l &= \frac{\partial x'^k}{\partial x^i}\frac{\partial x^j}{\partial x'^l} A^i_j + \frac{\partial x'^k}{\partial x^i}\frac{\partial x^j}{\partial x'^l} B^i_j \\ &= \frac{\partial x'^k}{\partial x^i}\frac{\partial x^j}{\partial x'^l}(A^i_j + B^i_j). \end{aligned}$$

If you compare this last expression to the expression for the transformation of the tensor components C^i_j to the primed coordinate system

$$C'^k_l = \frac{\partial x'^k}{\partial x^i}\frac{\partial x^j}{\partial x'^l}C^i_j,$$

you'll see that the addition of A^i_j and B^i_j does produce an object C^i_j that meets the transformation requirements for a tensor.

Subtraction of tensors is equally straightforward; you simply subtract the corresponding components rather than adding them:

$$\begin{aligned} C_{ij} &= A_{ij} - B_{ij}, \\ C^{ij} &= A^{ij} - B^{ij}, \\ C^i_j &= A^i_j - B^i_j, \end{aligned} \tag{5.10}$$

and the result of tensor subtraction is also a tensor, as you can see in the problems at the end of this chapter.

5.4 Tensor multiplication

As described in Chapter 2, there are several different ways to multiply vectors – the scalar (dot) product and vector (cross) product both take two vectors as inputs and produce a result that depends on the magnitudes and directions of those two vectors. Not mentioned in that chapter was another form of vector product called the "outer" product between a column vector ($\vec{A}$) and a row vector ($\vec{B}$), which operates like this:

$$\vec{A}\otimes\vec{B} = \begin{pmatrix} A_1 \\ A_2 \\ A_3 \end{pmatrix}(B_1 B_2 B_3) = \begin{pmatrix} A_1B_1 & A_1B_2 & A_1B_3 \\ A_2B_1 & A_2B_2 & A_2B_3 \\ A_3B_1 & A_3B_2 & A_3B_3 \end{pmatrix}.$$

Note that the outer product of two rank-1 tensors (vectors) is a rank-2 tensor, formed simply by multiplying the individual components of the two vectors. The outer product is indicated with the $\otimes$ symbol in some texts; others just write the two vectors or tensors next to one another, such as $A^iB^j = C^{ij}$.

The outer-product operation may also be performed on higher-order tensors:

$$A^i_j B^k_{lm} = C^{ik}_{jlm}.$$

In this case, the outer product of a rank-2 tensor and a rank-3 tensor is a rank-5 tensor. This illustrates the fact that the covariant rank of the outer-product tensor is the sum of the covariant ranks of the input tensors, and the contravariant rank of the outer-product tensor is the sum of the contravariant ranks of the input tensors.

The result of the outer-product operation is easily shown to be a tensor by considering how tensors $\vec{\vec{A}}$, $\vec{\vec{B}}$, and $\vec{\vec{C}}$ transform from the unprimed to the primed coordinate system. The transform of tensors $\vec{\vec{A}}$ and $\vec{\vec{B}}$ is given by

$$A'^{n}_{o} = \frac{\partial x'^{n}}{\partial x^{i}} \frac{\partial x^{j}}{\partial x'^{o}} A^{i}_{j},$$

$$B'^{p}_{qr} = \frac{\partial x'^{p}}{\partial x^{k}} \frac{\partial x^{l}}{\partial x'^{q}} \frac{\partial x^{m}}{\partial x'^{r}} B^{k}_{lm}.$$

Multiplying these expressions gives

$$\begin{aligned} A'^{n}_{o} B'^{p}_{qr} &= \frac{\partial x'^{n}}{\partial x^{i}} \frac{\partial x^{j}}{\partial x'^{o}} A^{i}_{j} \frac{\partial x'^{p}}{\partial x^{k}} \frac{\partial x^{l}}{\partial x'^{q}} \frac{\partial x^{m}}{\partial x'^{r}} B^{k}_{lm} \\ &= \frac{\partial x'^{n}}{\partial x^{i}} \frac{\partial x^{j}}{\partial x'^{o}} \frac{\partial x'^{p}}{\partial x^{k}} \frac{\partial x^{l}}{\partial x'^{q}} \frac{\partial x^{m}}{\partial x'^{r}} A^{i}_{j} B^{k}_{lm}. \end{aligned}$$

So if $A^{i}_{j} B^{k}_{lm} = C^{ik}_{jlm}$ and $A'^{n}_{o} B'^{p}_{qr} = C'^{np}_{oqr}$, then

$$C'^{np}_{oqr} = \frac{\partial x'^{n}}{\partial x^{i}} \frac{\partial x^{j}}{\partial x'^{o}} \frac{\partial x'^{p}}{\partial x^{k}} \frac{\partial x^{l}}{\partial x'^{q}} \frac{\partial x^{m}}{\partial x'^{r}} C^{ik}_{jlm}, \tag{5.11}$$

and the result of the outer product operation does indeed meet the transformation requirements for a tensor.

Another way to multiply tensors is called the "inner product," which you can think of as a generalization of the scalar or dot product discussed in Section 2.1. As described in that section, the dot product between two vectors produces a scalar result, so you might expect the inner product between two tensors to produce a tensor of lower rank. That's exactly right, but to understand how it happens, you first need to understand the process of tensor contraction.

To contract a tensor, simply set one contravariant index equal to a covariant index (or vice versa) and then sum over the repeated index. This leads to a tensor with a rank that is two less than the rank of the tensor with which you started.

To see how this works in practice, consider the rank-4 tensor C^{ij}_{kl}. To contract this tensor in the second and third indices, set the index k equal to the index j, resulting in

$$C^{ij}_{jl} = C^{i1}_{1l} + C^{i2}_{2l} + C^{i3}_{3l} = D^{i}_{l},$$

assuming that the indices j and k run from 1 to 3. Note that the rank is reduced by two because you made one index the same as another (reducing the rank by one) and then you summed over that index (reducing the rank by one more). Note also that contraction produces another tensor only when the two indices that are made equal are in different positions (one superscript and one subscript).

The reason for this becomes clear if you consider the contraction of the tensor that resulted from the outer-product operation in Eq. 5.11. Contracting this tensor in the first and fourth indices by setting q equal to n gives

$$\begin{aligned} C'^{np}_{onr} &= \frac{\partial x'^n}{\partial x^i}\frac{\partial x^j}{\partial x'^o}\frac{\partial x'^p}{\partial x^k}\frac{\partial x^l}{\partial x'^n}\frac{\partial x^m}{\partial x'^r}C^{ik}_{jlm} \\ &= \frac{\partial x'^n}{\partial x^i}\frac{\partial x^l}{\partial x'^n}\frac{\partial x^j}{\partial x'^o}\frac{\partial x'^p}{\partial x^k}\frac{\partial x^m}{\partial x'^r}C^{ik}_{jlm} \\ &= \frac{\partial x^l}{\partial x^i}\frac{\partial x^j}{\partial x'^o}\frac{\partial x'^p}{\partial x^k}\frac{\partial x^m}{\partial x'^r}C^{ik}_{jlm}. \end{aligned}$$

But the derivative $\frac{\partial x^l}{\partial x^i}$ involves only coordinates in the same (unprimed) system, and coordinates within the same system must be independent of one another. Hence this derivative must equal zero unless $l = i$, in which case it must equal one. This is most easily expressed using the Kronecker Delta function, defined by

$$\delta^i_j = \begin{cases} 1 & i = j \\ 0 & i \neq j \end{cases}.$$

Thus

$$\begin{aligned} C'^{np}_{onr} &= \delta^i_l \frac{\partial x^j}{\partial x'^o}\frac{\partial x'^p}{\partial x^k}\frac{\partial x^m}{\partial x'^r}C^{ik}_{jlm} \\ &= \frac{\partial x^j}{\partial x'^o}\frac{\partial x'^p}{\partial x^k}\frac{\partial x^m}{\partial x'^r}C^{ik}_{jim}, \end{aligned}$$

which is a tensor of rank 3, as expected. But note that this reduction from 5 to 3 in rank required that two of the partial derivatives combine to produce the delta function, which then invoked the summation process. That derivative combination only works if one of the contracted indices is a superscript and the other a subscript.

In this last example, the contraction was performed on a tensor that was the result of an outer product. That two-step process (outer-product multiplication followed by contraction) is called the "inner product" of two tensors. So if you start with two vectors (tensors of rank 1), form their outer product (producing a tensor of rank 2), and then contract the result, you end up with a tensor of rank zero – a scalar. This illustrates why the inner-product process can be considered to be a generalization of the dot product between two vectors.

5.5 Metric tensor

As you think about contravariant and covariant components of vectors and tensors, you should not lose sight of the fact that these components exist only when you've selected a coordinate system. And why do you need a coordinate system? Because coordinate systems "arithmetize" space – that is, they give you a way of applying the rules of arithmetic to objects that exist in the space in which you're working. That space may be the three-dimension space of everyday experience, or the four-dimension spacetime of Einstein, or any other space you can imagine. The coordinate system you apply may have straight axes that intersect at right angles, or the axes may be curved and intersect at any angle of your choosing.

However you choose to arithmetize a space, there is one tensor that allows you to define fundamental quantities such as lengths and angles in a consistent manner at different locations. That tensor, the one that "provides the metric" for a given coordinate system in the space of interest, is called the fundamental or metric tensor. The lower-case letter "g" has become the standard symbol for the metric tensor, which you may see written as $\vec{\vec{g}}$ or **g**. The metric tensor has contravariant components g^{ij} and covariant components g_{ij}.

To understand the role of the metric tensor, consider two points separated by an infinitesimal distance ds. If the vector $d\vec{r}$ extends from one point to the other, then the square of the differential length element may be written as $ds^2 = d\vec{r} \circ d\vec{r}$. The vector $d\vec{r}$ may be written using contravariant components and coordinate basis vectors ($\vec{e}_i$) as

$$d\vec{r} = \vec{e}_i dx^i,$$

or using covariant components and dual basis vectors ($\vec{e}_i$) as

$$d\vec{r} = \vec{e}^i dx_i.$$

Since ds^2 involves the dot product of $d\vec{r}$ with itself, you have the option of using the contravariant components dx^i on both sides of the dot:

$$\begin{aligned} ds^2 = d\vec{r} \circ d\vec{r} &= \vec{e}_i dx^i \circ \vec{e}_j dx^j \\ &= (\vec{e}_i \circ \vec{e}_j) dx^i dx^j \\ &= g_{ij} dx^i dx^j, \end{aligned}$$

where g_{ij} represents the covariant components of the metric tensor. Alternatively, you may use the covariant components dx_i on both sides of the dot:

$$ds^2 = d\vec{r} \circ d\vec{r} = \vec{e}^i dx_i \circ \vec{e}^j dx_j$$
$$= (\vec{e}^i \circ \vec{e}^j) dx_i dx_j$$
$$= g^{ij} dx_i dx_j,$$

where g^{ij} represents contravariant components of the metric tensor. A third option is to use contravariant components on one side of the dot and covariant components on the other:

$$ds^2 = \vec{e}_i dx^i \circ \vec{e}^j dx_j$$
$$= (\vec{e}_i \circ \vec{e}^j) dx^i dx_j$$
$$= dx^i dx_j.$$

Note that in this case no metric tensor is needed, since the definition of dual basis vectors ensures that $\vec{e}_i \circ \vec{e}^j$ equals one if $i = j$ and zero if $i \neq j$.

Whether ds^2 is written as $g_{ij} dx^i dx^j$, $g^{ij} dx_i dx_j$, or $dx^i dx_j$, you can be sure of one thing: the distance between two points must be the same no matter which coordinate system you employ, whether you use contravariant, covariant, or mixed components. Hence it must be the job of the metric tensor $\vec{\vec{g}}$ and its components g^{ij} and g_{ij} to turn the product of incremental coordinate changes expressed in either contravariant or covariant components into the invariant distance between points. This is the rationale behind the statement that the metric tensor "provides the geometry" of the space.

The geometry of vectors entails use of lengths and angles, so it's useful to understand the role of the metric tensor in defining the length of a vector such as $\vec{A}$ and the angle between two vectors $\vec{A}$ and $\vec{B}$. Just as the incremental distance ds can be found by dotting the separation vector $d\vec{r}$ into itself, the length of vector $\vec{A}$ can be found from $\vec{A} \circ \vec{A}$. And there's more than one way to do that.

One option is to use only the contravariant components of $\vec{A}$:

$$|\vec{A}| = \sqrt{\vec{A} \circ \vec{A}} = \sqrt{A^i \vec{e}_i \circ A^j \vec{e}_j}$$
$$= \sqrt{(\vec{e}_i \circ \vec{e}_j) A^i A^j} = \sqrt{g_{ij} A^i A^j}.$$

Another option is to use only covariant components:

$$|\vec{A}| = \sqrt{\vec{A} \circ \vec{A}} = \sqrt{A_i \vec{e}^i \circ A_j \vec{e}^j}$$
$$= \sqrt{(\vec{e}^i \circ \vec{e}^j) A_i A_j} = \sqrt{g^{ij} A_i A_j}.$$

And the final option is to use mixed components:

$$|\vec{A}| = \sqrt{\vec{A} \circ \vec{A}} = \sqrt{A^i \vec{e}_i \circ A_j \vec{e}^{\,j}}$$
$$= \sqrt{(\vec{e}_i \circ \vec{e}^{\,j}) A^i A_j} = \sqrt{A^i A_j}.$$

As in the case of $d\vec{r}$, the metric tensor ensures that the length of vector $\vec{A}$ is invariant.

To understand the role of the metric tensor in providing a consistent definition of angles, consider the dot product $\vec{A} \circ \vec{B}$. Once again, there are alternative ways of writing this product, and this means that the angle between $\vec{A}$ and $\vec{B}$ can be written in the following equivalent ways:

$$\cos\theta = \frac{\vec{A} \circ \vec{B}}{|\vec{A}||\vec{B}|}$$
$$= \frac{g_{ij} A^i B^j}{\sqrt{g_{ij} A^i A^j}\sqrt{g_{ij} B^i B^j}}$$
$$= \frac{A_i B^j}{\sqrt{A_i A^i}\sqrt{B_i B^i}}$$
$$= \frac{g^{ij} A_i B_j}{\sqrt{g^{ij} A_i A_j}\sqrt{g^{ij} B_i B_j}}.$$

This explains why you're likely to run into the statement that the metric tensor "provides a dot product" for a space – if you know how to find the dot product, you can define lengths and angles.

To see the tensor nature of the metric tensor, consider the transformation of the contravariant components of the incremental separation vector $d\vec{r}$:

$$dx'^i = \frac{\partial x'^i}{\partial x^j} dx^j.$$

This means that the square of the incremental length (ds^2) becomes:

$$ds^2 = \left[\frac{\partial x'^1}{\partial x^1}\frac{\partial x'^1}{\partial x^1} + \frac{\partial x'^2}{\partial x^1}\frac{\partial x'^2}{\partial x^1} + \frac{\partial x'^3}{\partial x^1}\frac{\partial x'^3}{\partial x^1}\right] dx^1 dx^1$$
$$+ \left[\frac{\partial x'^1}{\partial x^2}\frac{\partial x'^1}{\partial x^2} + \frac{\partial x'^2}{\partial x^2}\frac{\partial x'^2}{\partial x^2} + \frac{\partial x'^3}{\partial x^2}\frac{\partial x'^3}{\partial x^2}\right] dx^2 dx^2$$
$$+ \left[\frac{\partial x'^1}{\partial x^3}\frac{\partial x'^1}{\partial x^3} + \frac{\partial x'^2}{\partial x^3}\frac{\partial x'^2}{\partial x^3} + \frac{\partial x'^3}{\partial x^3}\frac{\partial x'^3}{\partial x^3}\right] dx^3 dx^3$$
$$+ \left[\frac{\partial x'^1}{\partial x^1}\frac{\partial x'^1}{\partial x^2} + \frac{\partial x'^2}{\partial x^1}\frac{\partial x'^2}{\partial x^2} + \frac{\partial x'^3}{\partial x^1}\frac{\partial x'^3}{\partial x^2}\right] dx^1 dx^2$$

$$+\left[\frac{\partial x'^1}{\partial x^2}\frac{\partial x'^1}{\partial x^1}+\frac{\partial x'^2}{\partial x^2}\frac{\partial x'^2}{\partial x^1}+\frac{\partial x'^3}{\partial x^2}\frac{\partial x'^3}{\partial x^1}\right]dx^2dx^1$$

$$+\left[\frac{\partial x'^1}{\partial x^1}\frac{\partial x'^1}{\partial x^3}+\frac{\partial x'^2}{\partial x^1}\frac{\partial x'^2}{\partial x^3}+\frac{\partial x'^3}{\partial x^1}\frac{\partial x'^3}{\partial x^3}\right]dx^1dx^3$$

$$+\left[\frac{\partial x'^1}{\partial x^3}\frac{\partial x'^1}{\partial x^1}+\frac{\partial x'^2}{\partial x^3}\frac{\partial x'^2}{\partial x^1}+\frac{\partial x'^3}{\partial x^3}\frac{\partial x'^3}{\partial x^1}\right]dx^3dx^1$$

$$+\left[\frac{\partial x'^1}{\partial x^2}\frac{\partial x'^1}{\partial x^3}+\frac{\partial x'^2}{\partial x^2}\frac{\partial x'^2}{\partial x^3}+\frac{\partial x'^3}{\partial x^2}\frac{\partial x'^3}{\partial x^3}\right]dx^2dx^3$$

$$+\left[\frac{\partial x'^1}{\partial x^3}\frac{\partial x'^1}{\partial x^2}+\frac{\partial x'^2}{\partial x^3}\frac{\partial x'^2}{\partial x^2}+\frac{\partial x'^3}{\partial x^3}\frac{\partial x'^3}{\partial x^2}\right]dx^3dx^2. \tag{5.12}$$

This daunting expression becomes far more tractable if you realize that each bracketed term involves the sum of the partial derivatives of each of the transformed coordinates (x'^1, x'^2, and x'^3) taken with respect to two of the original coordinates (x^1, x^2, and x^3). More specifically, each of the three terms within each bracket is a product of the components of the basis vectors tangent to the original axes (recall that $\frac{\partial x'^1}{\partial x^i}$, $\frac{\partial x'^2}{\partial x^i}$, and $\frac{\partial x'^3}{\partial x^i}$ are the components in the transformed coordinate system of the basis vector tangent to the ith original axis).

If you assign the bracketed terms to the variable g with two subscripts denoting the axes with respect to which the derivatives are taken, you will have

$$g_{11}=\left[\frac{\partial x'^1}{\partial x^1}\frac{\partial x'^1}{\partial x^1}+\frac{\partial x'^2}{\partial x^1}\frac{\partial x'^2}{\partial x^1}+\frac{\partial x'^3}{\partial x^1}\frac{\partial x'^3}{\partial x^1}\right],$$

$$g_{22}=\left[\frac{\partial x'^1}{\partial x^2}\frac{\partial x'^1}{\partial x^2}+\frac{\partial x'^2}{\partial x^2}\frac{\partial x'^2}{\partial x^2}+\frac{\partial x'^3}{\partial x^2}\frac{\partial x'^3}{\partial x^2}\right],$$

$$g_{33}=\left[\frac{\partial x'^1}{\partial x^3}\frac{\partial x'^1}{\partial x^3}+\frac{\partial x'^2}{\partial x^3}\frac{\partial x'^2}{\partial x^3}+\frac{\partial x'^3}{\partial x^3}\frac{\partial x'^3}{\partial x^3}\right],$$

$$g_{12}=\left[\frac{\partial x'^1}{\partial x^1}\frac{\partial x'^1}{\partial x^2}+\frac{\partial x'^2}{\partial x^1}\frac{\partial x'^2}{\partial x^2}+\frac{\partial x'^3}{\partial x^1}\frac{\partial x'^3}{\partial x^2}\right],$$

$$g_{13}=\left[\frac{\partial x'^1}{\partial x^1}\frac{\partial x'^1}{\partial x^3}+\frac{\partial x'^2}{\partial x^1}\frac{\partial x'^2}{\partial x^3}+\frac{\partial x'^3}{\partial x^1}\frac{\partial x'^3}{\partial x^3}\right],$$

$$g_{23}=\left[\frac{\partial x'^1}{\partial x^2}\frac{\partial x'^1}{\partial x^3}+\frac{\partial x'^2}{\partial x^2}\frac{\partial x'^2}{\partial x^3}+\frac{\partial x'^3}{\partial x^2}\frac{\partial x'^3}{\partial x^3}\right],$$

and since the order of multiplication is irrelevant, $g_{21} = g_{12}$, $g_{31} = g_{13}$, and $g_{32} = g_{23}$. Substituting these into Eq. 5.12, the expression for ds^2 becomes

$$ds^2 = g_{11}dx^1dx^1 + g_{22}dx^2dx^2 + g_{33}dx^3dx^3 + g_{12}dx^1dx^2 + g_{21}dx^2dx^1 + g_{13}dx^1dx^3 + g_{31}dx^3dx^1 + g_{23}dx^2dx^3 + g_{32}dx^3dx^2.$$

This can be further simplified using index notation and the summation convention:

$$ds^2 = g_{ij}dx^i dx^j. \tag{5.13}$$

The g_{ij} term in this equation meets all the requirements of a second-rank tensor, but it's not just any tensor. Because it relates the coordinate differentials in various directions to a quantity that is invariant across all coordinate transformations, it's no wonder that this tensor is called the metric or fundamental tensor.

To understand what's so fundamental about this tensor, recall that the partial derivatives that make up the elements of g_{ij} also represent the components of the basis vectors tangent to the original coordinate axes:

$$\begin{aligned}\vec{e}_1 &= \left(\frac{\partial x'^1}{\partial x^1}, \frac{\partial x'^2}{\partial x^1}, \frac{\partial x'^3}{\partial x^1}\right),\\ \vec{e}_2 &= \left(\frac{\partial x'^1}{\partial x^2}, \frac{\partial x'^2}{\partial x^2}, \frac{\partial x'^3}{\partial x^2}\right),\\ \vec{e}_3 &= \left(\frac{\partial x'^1}{\partial x^3}, \frac{\partial x'^2}{\partial x^3}, \frac{\partial x'^3}{\partial x^3}\right).\end{aligned} \tag{5.14}$$

And since

$$g_{ij} = \left[\frac{\partial x'^1}{\partial x^i}\frac{\partial x'^1}{\partial x^j} + \frac{\partial x'^2}{\partial x^i}\frac{\partial x'^2}{\partial x^j} + \frac{\partial x'^3}{\partial x^i}\frac{\partial x'^3}{\partial x^j}\right], \tag{5.15}$$

another way to represent the metric tensor is $g_{ij} = \vec{e}_i \circ \vec{e}_j$ (the inner product of the basis vectors tangent to the coordinate axes). Since the inner product involves the projection of one vector onto the direction of another and scales as the length of those two vectors, the elements of g_{ij} specify the relationships between the coordinate axes. Those relationships are determined by the shape of the coordinate space.

The nature of the metric tensor can be readily understood by considering a transformation from spherical polar (r, θ, ϕ) to Cartesian (x, y, z) coordinates. In this case

$$
\begin{aligned}
x'^1 &= x = r sin(\theta) cos(\phi) = x^1 sin(x^2) cos(x^3), \\
x'^2 &= y = r sin(\theta) sin(\phi) = x^1 sin(x^2) sin(x^3), \\
x'^3 &= z = r cos(\theta) = x^1 cos(x^2),
\end{aligned} \tag{5.16}
$$

and the partial derivatives appearing in the elements of the metric tensor are

$$
\begin{aligned}
\frac{\partial x'^1}{\partial x^1} &= sin(x^2) cos(x^3) = sin(\theta) cos(\phi), \\
\frac{\partial x'^1}{\partial x^2} &= x^1 cos(x^2) cos(x^3) = r cos(\theta) cos(\phi), \\
\frac{\partial x'^2}{\partial x^1} &= sin(x^2) sin(x^3) = sin(\theta) sin(\phi), \\
\frac{\partial x'^2}{\partial x^2} &= x^1 cos(x^2) sin(x^3) = r cos(\theta) sin(\phi), \\
\frac{\partial x'^3}{\partial x^1} &= cos(x^2) = cos(\theta), \\
\frac{\partial x'^3}{\partial x^2} &= -x^1 sin(x^2) = -r sin(\theta),
\end{aligned}
$$

and

$$
\begin{aligned}
\frac{\partial x'^1}{\partial x^3} &= -x^1 sin(x^2) sin(x^3) = -r sin(\theta) sin(\phi), \\
\frac{\partial x'^2}{\partial x^3} &= x^1 sin(x^2) cos(x^3) = r sin(\theta) cos(\phi), \\
\frac{\partial x'^3}{\partial x^3} &= 0.
\end{aligned}
$$

Inserting these values into the expression for g_{ij} (Eq. 5.15) gives the diagonal terms:[3]

$$
\begin{aligned}
g_{11} &= \left[\frac{\partial x'^1}{\partial x^1}\frac{\partial x'^1}{\partial x^1} + \frac{\partial x'^2}{\partial x^1}\frac{\partial x'^2}{\partial x^1} + \frac{\partial x'^3}{\partial x^1}\frac{\partial x'^3}{\partial x^1}\right] = 1, \\
g_{22} &= \left[\frac{\partial x'^1}{\partial x^2}\frac{\partial x'^1}{\partial x^2} + \frac{\partial x'^2}{\partial x^2}\frac{\partial x'^2}{\partial x^2} + \frac{\partial x'^3}{\partial x^2}\frac{\partial x'^3}{\partial x^2}\right] = r^2, \\
g_{33} &= \left[\frac{\partial x'^1}{\partial x^3}\frac{\partial x'^1}{\partial x^3} + \frac{\partial x'^2}{\partial x^3}\frac{\partial x'^2}{\partial x^3} + \frac{\partial x'^3}{\partial x^3}\frac{\partial x'^3}{\partial x^3}\right] = r^2 sin^2(\theta).
\end{aligned}
$$

[3] If you don't see how to get these results, you can find more detail in the problems at the end of this chapter and in the on-line solutions.

The off-diagonal terms are

$$g_{12} = \left[\frac{\partial x'^1}{\partial x^1} \frac{\partial x'^1}{\partial x^2} + \frac{\partial x'^2}{\partial x^1} \frac{\partial x'^2}{\partial x^2} + \frac{\partial x'^3}{\partial x^1} \frac{\partial x'^3}{\partial x^2} \right] = 0,$$

$$g_{13} = \left[\frac{\partial x'^1}{\partial x^1} \frac{\partial x'^1}{\partial x^3} + \frac{\partial x'^2}{\partial x^1} \frac{\partial x'^2}{\partial x^3} + \frac{\partial x'^3}{\partial x^1} \frac{\partial x'^3}{\partial x^3} \right] = 0,$$

$$g_{23} = \left[\frac{\partial x'^1}{\partial x^2} \frac{\partial x'^1}{\partial x^3} + \frac{\partial x'^2}{\partial x^2} \frac{\partial x'^2}{\partial x^3} + \frac{\partial x'^3}{\partial x^2} \frac{\partial x'^3}{\partial x^3} \right] = 0.$$

Thus the metric tensor for spherical polar coordinates is

$$g_{ij} = \begin{bmatrix} g_{11} & g_{12} & g_{13} \\ g_{21} & g_{22} & g_{23} \\ g_{31} & g_{32} & g_{33} \end{bmatrix} = \begin{bmatrix} 1 & 0 & 0 \\ 0 & r^2 & 0 \\ 0 & 0 & r^2 sin^2(\theta) \end{bmatrix}. \tag{5.17}$$

A careful look at the metric tensor can tell you something about the coordinate system you're dealing with. For example, the fact that all off-diagonal elements are zero in this case tells you that spherical polar coordinate axes, while curved, are orthogonal (that is, the lines of increasing r, θ, and ϕ intersect at right angles). Furthermore, by inserting these values into Eq. 5.13, you'll have

$$ds^2 = dr^2 + r^2 d\theta^2 + r^2 sin^2\theta d\phi^2. \tag{5.18}$$

This expression makes it clear that the elements of the metric tensor tell you how to turn an incremental change in r, θ, or ϕ into a change in distance. For example, the factor of one in front of the dr^2 term means that a change in r is already a distance. But a change in zenith angle (θ) must be multiplied by a factor of r to turn it into a distance. And the distance corresponding to a change in the azimuthal angle ϕ depends on both the zenith angle (hence the $sin(\theta)$ term in g_{33}) as well as the distance from the origin (hence the r term in g_{33}).

Other coordinate systems require other factors to convert each change in a coordinate value to a distance, and those factors always appear in the metric tensor for that system. For orthogonal coordinate systems, the square roots of the diagonal elements of the metric tensor ($\sqrt{g_{11}}$, $\sqrt{g_{22}}$, and $\sqrt{g_{33}}$) are called the "scale factors" (h_1, h_2, and h_3) of the coordinate system. Thus the scale factors for spherical polar coordinates are $h_1 = \sqrt{g_{11}} = 1$, $h_2 = \sqrt{g_{22}} = r$, and $h_3 = \sqrt{g_{33}} = r \sin\theta$.

Once you're familiar with the metric tensor and scale factors, you can easily find the differential operators gradient, divergence, curl, and Laplacian in any orthogonal coordinate system (curvilinear or rectangular). For example, the gradient is given by

$$\vec{\nabla}\phi = \frac{1}{h_1}\frac{\partial\phi}{\partial x^1}\hat{e}_1 + \frac{1}{h_2}\frac{\partial\phi}{\partial x^2}\hat{e}_2 + \frac{1}{h_3}\frac{\partial\phi}{\partial x^3}\hat{e}_3,$$

and the divergence may be written as

$$\vec{\nabla}\circ\vec{A} = \frac{1}{h_1h_2h_3}\left[\frac{\partial}{\partial x^1}(h_2h_3A_1) + \frac{\partial}{\partial x^2}(h_1h_3A_2) + \frac{\partial}{\partial x^3}(h_1h_2A_3)\right].$$

The curl is given by

$$\vec{\nabla}\times\vec{A} = \frac{1}{h_1h_2h_3}\begin{vmatrix} h_1\hat{e}_1 & h_2\hat{e}_2 & h_3\hat{e}_3 \\ \frac{\partial}{\partial x^1} & \frac{\partial}{\partial x^2} & \frac{\partial}{\partial x^3} \\ h_1A_1 & h_2A_2 & h_3A_3 \end{vmatrix},$$

which expands to

$$\vec{\nabla}\times\vec{A} = \frac{1}{h_1h_2h_3}\left[\left(\frac{\partial h_3A_3}{\partial x^2} - \frac{\partial h_2A_2}{\partial x^3}\right)h_1\hat{e}_1 + \left(\frac{\partial h_1A_1}{\partial x^3} - \frac{\partial h_3A_3}{\partial x^1}\right)h_2\hat{e}_2 + \left(\frac{\partial h_2A_2}{\partial x^1} - \frac{\partial h_1A_1}{\partial x^2}\right)h_3\hat{e}_3\right].$$

The Laplacian can be found as

$$\nabla^2\phi = \frac{1}{h_1h_2h_3}\left[\frac{\partial}{\partial x^1}\left(\frac{h_2h_3}{h_1}\frac{\partial\phi}{\partial x^1}\right) + \frac{\partial}{\partial x^2}\left(\frac{h_1h_3}{h_2}\frac{\partial\phi}{\partial x^2}\right) + \frac{\partial}{\partial x^3}\left(\frac{h_1h_2}{h_3}\frac{\partial\phi}{\partial x^3}\right)\right].$$

If you'd like to see some examples of how these expressions can be used, check out the problems at the end of this chapter and the on-line solutions.[4]

5.6 Index raising and lowering

One of the many useful functions of the metric tensor is to convert between the covariant and contravariant components of other tensors. Imagine that you're given the contravariant components and original basis vectors of a tensor and you wish to determine the covariant components. One approach is to use the techniques described in Chapter 4 (finding the dual basis vectors, performing parallel and perpendicular projections, and the like), but with the metric tensor, you have another option. You can use relations such as

$$g_{ij}A^j = A_i \tag{5.19}$$

to convert a contravariant index to a covariant one (thus "lowering" an index). Furthermore, if you wish to convert a covariant index to a contravariant index,

[4] You can find the derivation of these extremely handy equations in Boas' *Mathematical Methods in the Physical Sciences*, John Wiley and Sons, 2006.

you can use the inverse of g_{ij} (which is just g^{ij}) to perform operations like this:

$$g^{ij} B_i = B^j. \tag{5.20}$$

And this same process works for higher-order tensors:

$$\begin{aligned} g^{ij} A_{ik} &= A^j_k, \\ C^i_{jk} &= g_{js} C^{is}_k, \\ T^{ijk} &= g^{il} T^{jk}_l. \end{aligned} \tag{5.21}$$

5.7 Tensor derivatives and Christoffel symbols

In many applications, it's important to know how a vector field changes as you move from one location to another. For vectors expressed using Cartesian coordinates, taking the derivative of a vector is quite straightforward: you simply take the derivative of each of the vector's components. You can do that because the Cartesian basis vectors ($\hat{\imath}$, $\hat{\jmath}$, and $\hat{k}$) are everywhere constant in both magnitude and direction. That means you don't need to worry about the derivatives of the basis vectors. But as you've seen for spherical polar coordinates, the basis vectors ($\hat{r}$, $\hat{\theta}$, and $\hat{\phi}$) point in different directions as you move around the space, which means that when you take a spatial derivative of a vector expressed in these coordinates, you must also consider the derivatives of the basis vectors.

Thus if you have a vector $\vec{A}$ expressed in general coordinates x^1, x^2, x^3 with covariant basis vectors $\vec{e}_1$, $\vec{e}_2$, and $\vec{e}_3$ as

$$\vec{A} = A^1\vec{e}_1 + A^2\vec{e}_2 + A^3\vec{e}_3,$$

the derivative of $\vec{A}$ with respect to coordinate x^1 is

$$\begin{aligned} \frac{\partial \vec{A}}{\partial x^1} &= \frac{\partial(A^1\vec{e}_1 + A^2\vec{e}_2 + A^3\vec{e}_3)}{\partial x^1} \\ &= \frac{\partial(A^i\vec{e}_i)}{\partial x^1} \\ &= \frac{\partial A^i}{\partial x^1}\vec{e}_1 + A^i\frac{\partial \vec{e}_i}{\partial x^1}. \end{aligned}$$

It's the second term in this equation that complicates the process of taking a derivative in coordinate systems in which the magnitude and/or direction of the basis vectors change as you move around the space. And as you might expect, similar terms appear when you take the derivatives of $\vec{A}$ with respect to the other coordinates. So if you want to evaluate the changes in vector fields expressed in non-orthonormal coordinates, you have to account for possible changes in the basis vectors. Properly accounting for those changes means that the result of the defferentiation process will retain the tensor characteristics of the original object.

Fortunately, there's a way to account for any change in the basis vectors and to ensure that the derivative of a tensor is another tensor. That process, called the "covariant derivative," is described in the next section of this chapter. But the process of covariant differentiation will make a lot more sense to you if you've first learned the meaning of the Christoffel symbols described in this section.

To understand Christoffel symbols, you should begin by realizing that the derivative of a basis vector will be another vector. Like any vector, that vector can be described as the weighted combination of the basis vectors at the point under consideration. Each Christoffel symbol, written as an uppercase Greek gamma (Γ), simply represents the weighting coefficient for one of the basis vectors. Hence the defining relationship for Christoffel symbols[5] is

$$\Gamma^k_{ij}\vec{e}_k = \frac{\partial \vec{e}_i}{\partial x^j}, \tag{5.22}$$

in which the index i specifies the basis vector for which the derivative is being taken, the index j denotes the coordinate being varied to induce this change in the ith basis vector, and the index k identifies the direction in which this component of the derivative points, as shown in Figure 5.1.

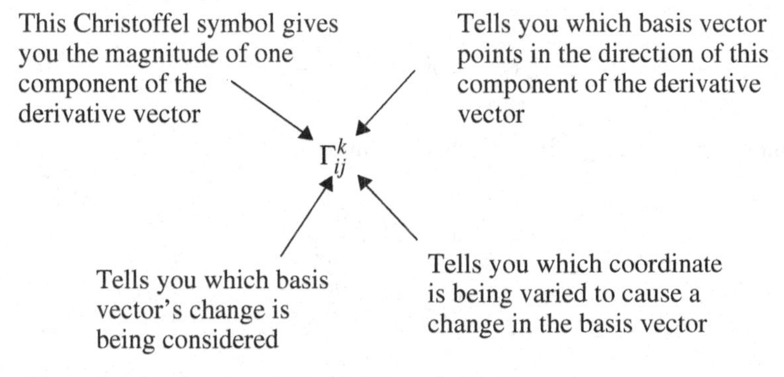

Figure 5.1 Explanation of Christoffel symbol indices.

[5] The Christoffel symbols written as Γ^k_{ij} are Christoffel symbols of the second kind; another form of Christoffel symbol (the "first kind") is described in most General Relativity texts.

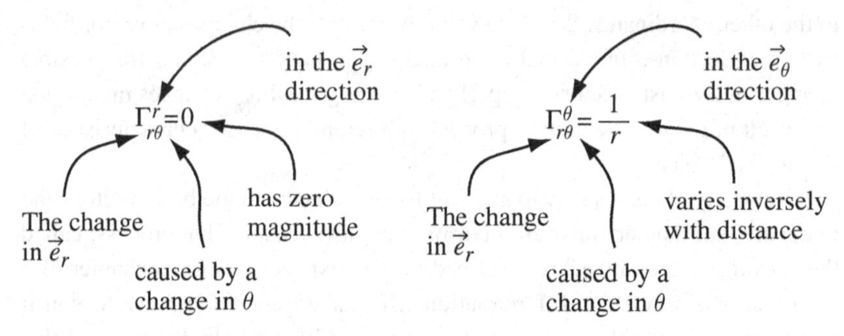

Figure 5.2 Example of Christoffel symbol indices.

Hence if you find two Christoffel symbols such as $\Gamma^r_{r\theta} = 0$ and $\Gamma^\theta_{r\theta} = \frac{1}{r}$, you know that

$$\frac{\partial \vec{e}_r}{\partial \theta} = 0\vec{e}_r + \frac{1}{r}\vec{e}_\theta,$$

which is further explained in Figure 5.2.

As this example illustrates, Christoffel symbols are really quite simple to understand once you know the code of their indices. Best of all, the values of these useful symbols are easy to determine if you know the elements of the metric tensor for the coordinate system in which you're working. It will take a bit of algebra to get to the relationship between Christoffel symbols and the metric tensor, but the result makes the trip worthwhile.

A good way to start is to form the dot product of the basis vector $\vec{e}^l$ with both sides of Eq. 5.22:

$$\Gamma^k_{ij}\vec{e}_k \circ \vec{e}^l = \vec{e}^l \circ \frac{\partial \vec{e}_i}{\partial x^j}.$$

Remembering that $\vec{e}_k \circ \vec{e}^l = \delta^l_k$, this becomes

$$\Gamma^k_{ij}\delta^l_k = \vec{e}^l \circ \frac{\partial \vec{e}_i}{\partial x^j},$$

$$\Gamma^l_{ij} = \vec{e}^l \circ \frac{\partial \vec{e}_i}{\partial x^j}.$$

Since the term $\frac{\partial \vec{e}_i}{\partial x^j}$ is the same as $\frac{\partial \vec{e}_j}{\partial x^i}$, this may be written as

$$\Gamma^l_{ij} = \frac{1}{2}\vec{e}^l \circ \frac{\partial \vec{e}_i}{\partial x^j} + \frac{1}{2}\vec{e}^l \circ \frac{\partial \vec{e}_j}{\partial x^i},$$

which seems rather pointless until you add nothing to it. Nothing, that is, in the following form:

$$\Gamma^l_{ij} = \frac{1}{2}\vec{e}^{\,l} \circ \frac{\partial \vec{e}_i}{\partial x^j} + \left(\frac{1}{2}g^{kl}\frac{\partial \vec{e}_k}{\partial x^j} \circ \vec{e}_i - \frac{1}{2}g^{kl}\frac{\partial \vec{e}_j}{\partial x^k} \circ \vec{e}_i\right)$$
$$+ \frac{1}{2}\vec{e}^{\,l} \circ \frac{\partial \vec{e}_j}{\partial x^i} + \left(\frac{1}{2}g^{kl}\frac{\partial \vec{e}_k}{\partial x^i} \circ \vec{e}_j - \frac{1}{2}g^{kl}\frac{\partial \vec{e}_i}{\partial x^k} \circ \vec{e}_j\right).$$

Note that the terms in parentheses on each line add to zero, so you haven't changed the quantity on the right side of the equation by adding these terms. It may look like things are getting worse, but the situation will become more clear once you've accomplished a few more bits of manipulation. The first bit is to realize that $\vec{e}^{\,l} = g^{kl}\vec{e}_k$, so the Christoffel symbol becomes

$$\Gamma^l_{ij} = \frac{1}{2}g^{kl}\vec{e}_k \circ \frac{\partial \vec{e}_i}{\partial x^j} + \left(\frac{1}{2}g^{kl}\frac{\partial \vec{e}_k}{\partial x^j} \circ \vec{e}_i - \frac{1}{2}g^{kl}\frac{\partial \vec{e}_j}{\partial x^k} \circ \vec{e}_i\right)$$
$$+ \frac{1}{2}g^{kl}\vec{e}_k \circ \frac{\partial \vec{e}_j}{\partial x^i} + \left(\frac{1}{2}g^{kl}\frac{\partial \vec{e}_k}{\partial x^i} \circ \vec{e}_j - \frac{1}{2}g^{kl}\frac{\partial \vec{e}_i}{\partial x^k} \circ \vec{e}_j\right).$$

Now it's just a matter pulling out the common factor of $\frac{1}{2}g^{kl}$ and grouping the terms by their sign:

$$\Gamma^l_{ij} = \frac{1}{2}g^{kl}\left[\left(\vec{e}_k \circ \frac{\partial \vec{e}_i}{\partial x^j} + \frac{\partial \vec{e}_k}{\partial x^j} \circ \vec{e}_i\right) + \left(\vec{e}_k \circ \frac{\partial \vec{e}_j}{\partial x^i} + \frac{\partial \vec{e}_k}{\partial x^i} \circ \vec{e}_j\right)\right.$$
$$\left. - \left(\frac{\partial \vec{e}_j}{\partial x^k} \circ \vec{e}_i + \frac{\partial \vec{e}_i}{\partial x^k} \circ \vec{e}_j\right)\right],$$

which may be further simplified if you recognize that

$$\vec{e}_k \circ \frac{\partial \vec{e}_i}{\partial x^j} + \frac{\partial \vec{e}_k}{\partial x^j} \circ \vec{e}_i = \frac{\partial(\vec{e}_k \circ \vec{e}_i)}{\partial x^j},$$
$$\vec{e}_k \circ \frac{\partial \vec{e}_j}{\partial x^i} + \frac{\partial \vec{e}_k}{\partial x^i} \circ \vec{e}_j = \frac{\partial(\vec{e}_j \circ \vec{e}_k)}{\partial x^i},$$
$$\vec{e}_i \circ \frac{\partial \vec{e}_j}{\partial x^k} + \frac{\partial \vec{e}_i}{\partial x^k} \circ \vec{e}_j = \frac{\partial(\vec{e}_i \circ \vec{e}_j)}{\partial x^k}.$$

So

$$\Gamma^l_{ij} = \frac{1}{2}g^{kl}\left[\frac{\partial(\vec{e}_k \circ \vec{e}_i)}{\partial x^j} + \frac{\partial(\vec{e}_j \circ \vec{e}_k)}{\partial x^i} - \frac{\partial(\vec{e}_i \circ \vec{e}_j)}{\partial x^k}\right].$$

But you know from the definition of the elements of the metric tensor that $\vec{e}_i \circ \vec{e}_k = g_{ik}$ and that $\vec{e}_i \circ \vec{e}_j = g_{ij}$, which means you can write

$$\Gamma^l_{ij} = \frac{1}{2}g^{kl}\left[\frac{\partial g_{ik}}{\partial x^j} + \frac{\partial g_{jk}}{\partial x^i} - \frac{\partial g_{ij}}{\partial x^k}\right]. \tag{5.23}$$

With this expression, finding the Christoffel symbols for any coordinate system for which you know the metric tensor is quite straightforward. And why is that worth doing? Simply because using the Christoffel symbols, you can take a derivative of vectors and tensors that accounts for changes in the basis vectors as well as changes in the components. This preserves the most important property of a tensor: invariance across coordinate systems. Such covariant derivatives are the subject of the next section, but before getting to that, you might want to consider an example of the Christoffel symbols for a familiar coordinate system.

Consider the cylindrical coordinates (r, ϕ, and z) described in Section 1.5. In this system, the square of the differential length element is related to the coordinate differentials by $ds^2 = dr^2 + r^2 d\phi^2 + dz^2$. Hence the covariant metric tensor may be represented by

$$g_{ij} = \begin{bmatrix} g_{11} & g_{12} & g_{13} \\ g_{21} & g_{22} & g_{23} \\ g_{31} & g_{32} & g_{33} \end{bmatrix} = \begin{bmatrix} 1 & 0 & 0 \\ 0 & r^2 & 0 \\ 0 & 0 & 1 \end{bmatrix},$$

which suggests that most of the Christoffel symbols will be zero in this case. You can verify that by taking the derivatives indicated in Eq. 5.23, beginning with $l = 1$, $i = 1$, and $j = 1$ (and don't forget that the summation convention means that you must sum over k):

$$\begin{aligned} \Gamma^1_{11} = &\frac{1}{2} g^{11} \left[\frac{\partial g_{11}}{\partial x^1} + \frac{\partial g_{11}}{\partial x^1} - \frac{\partial g_{11}}{\partial x^1} \right] \\ &+ \frac{1}{2} g^{21} \left[\frac{\partial g_{12}}{\partial x^1} + \frac{\partial g_{12}}{\partial x^1} - \frac{\partial g_{11}}{\partial x^2} \right] \\ &+ \frac{1}{2} g^{31} \left[\frac{\partial g_{13}}{\partial x^1} + \frac{\partial g_{13}}{\partial x^1} - \frac{\partial g_{11}}{\partial x^3} \right], \end{aligned}$$

and then using the relations $x^1 = r$, $x^2 = \phi$, and $x^3 = z$:

$$\begin{aligned} \Gamma^1_{11} = &\frac{1}{2} (1) \left[\frac{\partial (1)}{\partial r} + \frac{\partial (1)}{\partial r} - \frac{\partial (1)}{\partial r} \right] \\ &+ \frac{1}{2} (0) \left[\frac{\partial (0)}{\partial r} + \frac{\partial (0)}{\partial r} - \frac{\partial (1)}{\partial \phi} \right] \\ &+ \frac{1}{2} (0) \left[\frac{\partial (0)}{\partial r} + \frac{\partial (0)}{\partial r} - \frac{\partial (1)}{\partial z} \right] = 0. \end{aligned}$$

OK, that one was pretty boring, as are most of the others in this case. But have a go at the Christoffel symbol for $l = 1$, $i = 2$, and $j = 2$:

$$\Gamma^1_{22} = \frac{1}{2}g^{11}\left[\frac{\partial g_{21}}{\partial x^2} + \frac{\partial g_{21}}{\partial x^2} - \frac{\partial g_{22}}{\partial x^1}\right]$$
$$+ \frac{1}{2}g^{21}\left[\frac{\partial g_{22}}{\partial x^2} + \frac{\partial g_{22}}{\partial x^2} - \frac{\partial g_{22}}{\partial x^2}\right]$$
$$+ \frac{1}{2}g^{32}\left[\frac{\partial g_{23}}{\partial x^2} + \frac{\partial g_{23}}{\partial x^2} - \frac{\partial g_{22}}{\partial x^3}\right],$$

which is:

$$\Gamma^1_{22} = \frac{1}{2}(1)\left[\frac{\partial(0)}{\partial \phi} + \frac{\partial(0)}{\partial \phi} - \frac{\partial r^2}{\partial r}\right]$$
$$+ \frac{1}{2}(0)\left[\frac{\partial r^2}{\partial \phi} + \frac{\partial r^2}{\partial \phi} - \frac{\partial r^2}{\partial \phi}\right]$$
$$+ \frac{1}{2}(0)\left[\frac{\partial(0)}{\partial \phi} + \frac{\partial(0)}{\partial \phi} - \frac{\partial r^2}{\partial z}\right],$$

or

$$\Gamma^1_{22} = \frac{1}{2}(1)[0 + 0 - 2r] + 0 + 0 = -r.$$

Now you're getting somewhere. And exactly where is that? Just remember the meaning of a Christoffel symbol, and you'll see that this result means that the change in the covariant $\vec{\phi}$ basis vector as you move in the ϕ direction has a component in the $-\vec{r}$ direction that increases directly with distance from the origin.

A similar analysis shows that $\Gamma^2_{12} = \Gamma^2_{21} = 1/r$, which are the only other non-zero Christoffel symbols for the cylindrical coordinate system.[6] If you don't see how to get that result, take a look at the problems at the end of this chapter and the on-line solutions.

5.8 Covariant differentiation

With Christoffel symbols in hand, you have a way of differentiating a vector or higher-order tensor that includes the effect of changes (if any) in the magnitude and direction of the basis vectors used to expand that vector or tensor. This type of derivative is called the "covariant" derivative, and it finds application not only in the Euclidean space in which many engineering and physics problems are worked, but also in the curved Riemanian space of General Relativity.

In Euclidean space, two vectors at different locations may be compared and combined by dragging one of the vectors to the location of the other without

[6] Note that the symmetry of the metric tensor means that Christoffel symbols of this type are symmetric in the two lower indices.

changing its magnitude or its direction. If the vector is expanded using Cartesian coordinates, such "parallel transport" is accomplished simply by keeping each of its components the same (because the Cartesian basis vectors have the same magnitude and direction everywhere). But if the vector is expressed in non-Cartesian coordinates, the length and direction of the basis vectors may be different at the two locations. In such cases, the covariant derivative provides a means of parallel-transporting one of the vectors to the location of the other.

The situation is more complicated for curved spaces. You can find the details of the use of the covariant derivative in curved spaces in Chapter 6, but for now you can understand the role of the covariant derivative by considering a two-dimensional spherical surface embedded in a three-dimensional Euclidean space. Imagine a series of tangent planes just touching the sphere at each location, and picture a vector lying in one of those tangent planes. If that vector is moved to a different location on the sphere while holding its direction constant (as viewed in the larger three-dimensional space), it will not lie in the tangent plane at the new location (you can think of the vector as "sticking out" of the two-dimensional space of the sphere). In such cases, the covariant derivative serves to project the derivative of the vector into the tangent space of the sphere.

You should also note that the covariant differentiation process produces a result that retains the properties of a tensor, which means that the result transforms between coordinate systems according to the rules of tensor transformation.

To understand how the process of covariant differentiation works, consider the vector $\vec{A} = A^1\vec{e}_1 + A^2\vec{e}_2 + A^3\vec{e}_3$ and its derivatives

$$\begin{aligned}\frac{\partial \vec{A}}{\partial x^j} &= \frac{\partial(A^1\vec{e}_1 + A^2\vec{e}_2 + A^3\vec{e}_3)}{\partial x^j} \\ &= \frac{\partial(A^i\vec{e}_i)}{\partial x^j} \\ &= \frac{\partial A^i}{\partial x^j}\vec{e}_i + A^i\frac{\partial \vec{e}_i}{\partial x^j}.\end{aligned}$$

Now replace the partial derivative in the second term with the Christoffel-symbol definition (Eq. 5.22):

$$\frac{\partial \vec{A}}{\partial x^j} = \frac{\partial A^i}{\partial x^j}\vec{e}_i + A^i(\Gamma^k_{ij}\vec{e}_k).$$

Since the indices i and k in the second term are both dummy indices by the summation rule, you can switch them and then extract the common factor that is now the basis vector $\vec{e}_i$:

$$\frac{\partial \vec{A}}{\partial x^j} = \frac{\partial A^i}{\partial x^j}\vec{e}_i + A^k(\Gamma^i_{kj}\vec{e}_i)$$
$$= \left(\frac{\partial A^i}{\partial x^j} + A^k\Gamma^i_{kj}\right)\vec{e}_i.$$

The covariant derivative is defined as the combination of the two terms inside the parentheses. Common notation for the covariant derivative is to use a semicolon (;) in front of the index with respect to which the covariant derivative is being taken (j in this case). Thus you're likely to see the components of the covariant derivative defined as

$$A^i_{\ ;j} \equiv \frac{\partial A^i}{\partial x^j} + A^k\Gamma^i_{kj}. \tag{5.24}$$

A similar analysis leads to the covariant derivative of a vector expanded using covariant coefficients:

$$A_{i;j} \equiv \frac{\partial A_i}{\partial x^j} - A_k\Gamma^k_{ij}. \tag{5.25}$$

Note that the term involving Christoffel symbols is subtracted in this case.

To make the meaning of Eqs. 5.24 and 5.25 more explicit, consider the covariant derivative of vector $\vec{A}$ with respect to ϕ in cylindrical coordinates (so $x^1 = r$, $x^2 = \phi$, and $x^3 = z$). Setting $j = 2$ in Eq. 5.24 (since we're interested in the covariant derivative with respect to ϕ),

$$A^r_{\ ;\phi} = \frac{\partial A^r}{\partial \phi} + A^r\Gamma^r_{r\phi} + A^\phi\Gamma^r_{\phi\phi} + A^z\Gamma^r_{z\phi}$$
$$= \frac{\partial A^r}{\partial \phi} + 0 + A^\phi(-r) + 0,$$

which says that a change in the r-component of vector $\vec{A}$ caused by a change in ϕ is caused both by a change in A^r with ϕ and by a change in the basis vectors which causes a portion of $\vec{A}$ that was originally in the ϕ-direction to now point in the $-r$-direction. Likewise, for the change in A^ϕ as the value of ϕ is changed,

$$A^\phi_{\ ;\phi} = \frac{\partial A^\phi}{\partial \phi} + A^\phi\Gamma^\phi_{r\phi} + A^r\Gamma^\phi_{\phi\phi} + A^z\Gamma^\phi_{z\phi}$$
$$= \frac{\partial A^\phi}{\partial \phi} + A^r\left(\frac{1}{r}\right) + 0 + 0.$$

Thus

$$\frac{\partial \vec{A}}{\partial \phi} = \left(\frac{\partial A^r}{\partial \phi} - rA^\phi\right)\vec{e}_r + \left(\frac{\partial A^\phi}{\partial \phi} + \frac{1}{r}A^r\right)\vec{e}_\phi.$$

The process of covariant differentiation can also be applied to higher-order tensors. As you might expect, this simply requires the addition of a

Christoffel-symbol term for each contravariant index, and the subtraction of a Christoffel-symbol term for each covariant index. Hence

$$A^{ij}{}_{;k} = \frac{\partial A^{ij}}{\partial x^k} + A^{lj}\Gamma^i_{lk} + A^{il}\Gamma^j_{lk},$$

$$B_{ij\,;k} = \frac{\partial B_{ij}}{\partial x^k} - B_{lj}\Gamma^l_{ik} - B_{il}\Gamma^l_{jk},$$

$$C^i{}_{j\,;k} = \frac{\partial C^i{}_j}{\partial x^k} + C^l{}_j\Gamma^i_{lk} - C^i{}_l\Gamma^l_{jk}.$$

5.9 Vectors and one-forms

If you look up the subject of tensors in recently published physics texts, especially those dealing with General Relativity, you may be surprised to find little mention of contravariant and covariant components in favor of terms such as "covectors" and "one-forms." Have you wasted your time struggling to understand complicated concepts and terminology that have now become obsolete? I obviously don't think so, or I wouldn't have devoted so many pages to the developments of the last two chapters. Instead, I believe there's value in seeing the "traditional" presentation as well as the "modern" approach, because the differences arise from perspective rather than from the core concepts. But those different perspectives do lead to very different terminology, and the purpose of this section is to provide a short introduction to that terminology.

The first thing to understand is that the traditional approach tends to treat contravariant and covariant components as representations of the same object, whereas in the modern approach objects are classified either as "vectors" or as "one-forms" (also called "covectors"). In the modern terminology, vectors transform as contravariant quantities, and one-forms transform as covariant quantities. Quantities with dimension of length in the numerator (such as velocity, with units that include "meters per") fit naturally into the vector category; quantities with dimension of length in denominator (such as the gradient of a scalar field, with units that include "per meter") fit naturally into the one-form category.

In illustrations involving vectors and one-forms, vectors are represented as arrows and one-forms are represented as small sections of surfaces, as shown in Figure 5.3. As indicated in the figure, for vectors the angle of the arrow shows direction and the length of the arrow shows the magnitude. For one-forms, surfaces are aligned normal to the direction and the spacing between surfaces is inversely proportional to the magnitude. This means that vectors with greater magnitude are represented by longer arrows, while one-forms of greater magnitude are represented by closer spacing.

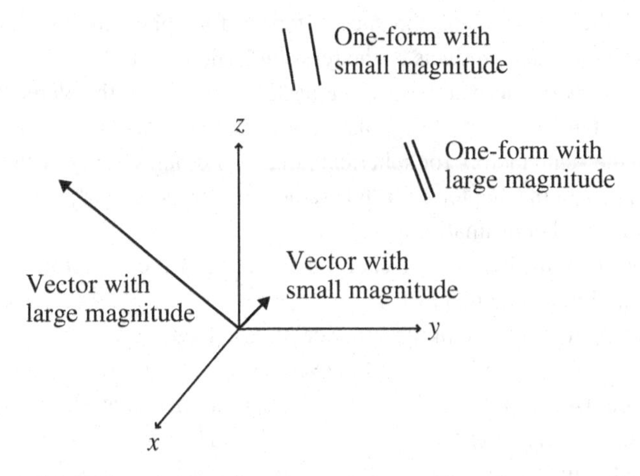

Figure 5.3 Representation of vectors as arrows and one-forms as surfaces.

As in the traditional approach, vectors (which utilize contravariant components) expand using original basis vectors, while one-forms (which utilize covariant components) expand using basis one-forms, which are equivalent to dual basis vectors in the traditional approach. That correspondence means that the product of a vector and a one-form is an invariant (a scalar), just as the multiplication of a contravariant and a covariant quantity produces a scalar without requiring the metric tensor. One very nice graphical interpretation of such products is that the resulting scalar is represented by the number of one-form surfaces through which the arrow of a vector passes.

Authors using the modern approach often place strong emphasis on vectors and one-forms as operators (or rules), so you're likely to encounter statements that vectors "take" one-forms and produce scalars, just as one-forms "take" vectors and produce scalars. Likewise, a higher-order tensor takes multiple vectors and/or one-forms and produces a scalar. From this perspective, the metric tensor is an operator that takes two vectors or two one-forms and produces their dot product, and the components of the metric tensor may be found by feeding it basis vectors or one-forms.

5.10 Chapter 5 problems

5.1 Show that the process of subtracting one tensor from another results in a quantity that is also a tensor.

5.2 Find the elements of the metric tensor for spherical coordinates by forming the dot products of the relevant basis vectors.

5.3 Show how the derivatives given after Eq. 5.16 lead to the elements of the metric tensor for spherical polar coordinates (Eq. 5.17).

5.4 Use the scale factors for spherical polar coordinates to verify the expressions given in Chapter 2 for the gradient, divergence, curl, and Laplacian in spherical coordinates.

5.5 Show that for cylindrical coordinates (r, ϕ, z) the Christoffel symbols Γ^2_{12} and Γ^2_{21} are equal to $1/r$.

5.6 Find g^{ij}, the inverse of the spherical metric tensor g_{ij}.

5.7 Use g^{ij} to raise the indices of the vector $A_i = (1, r^2 sin\theta, sin^2\theta)$.

5.8 On the two-dimensional surface of a sphere of radius R, the square of the differential length element is given by $ds^2 = R^2 d\theta^2 + R^2 sin^2\theta d\phi^2$. Find the metric tensor g_{ij} and its inverse g^{ij} for this case.

5.9 What are the Christoffel symbols for the 2-D spherical surface of Problem 5.8?

5.10 Show that the covariant derivative of the metric tensor equals zero.

6
Tensor applications

This chapter provides examples of how to apply the tensor concepts contained in Chapters 4 and 5, just as Chapter 3 provided examples of how to apply the vector concepts presented in Chapters 1 and 2. As in Chapter 3, the intent for this chapter is to include more detail about a small number of selected applications than can be included in the chapters in which tensor concepts are first presented.

The examples in this chapter come from the fields of Mechanics, Electromagnetics, and General Relativity. Of course, there's no way to comprehensively cover any significant portion of those fields in one chapter; these examples were chosen only to serve as representatives of the types of tensor application you're likely to encounter in those fields.

6.1 The inertia tensor

A very useful way to think of mass is this: mass is the characteristic of matter that resists acceleration. This means that it takes a force to change the velocity of any object with mass. You may find it helpful to think of moment of inertia as the rotational analog of mass. That is, moment of inertia is the characteristic of matter that resists angular acceleration, so it takes a torque to change the angular velocity of an object.

Many students find that rotational motion is easier to understand by keeping the relationships between translational and rotational quantities in mind. So where translational motion dealt with position (x), velocity ($\vec{v}$), and acceleration ($\vec{a}$), rotational motion has the analogous quantities of angle (θ), angular velocity ($\vec{\omega}$), and angular acceleration ($\vec{\alpha}$). There are rotational analogs for many other quantities; the translational quantities of force ($\vec{F}$), mass (m), and momentum ($\vec{p}$) have the rotational equivalents of torque ($\vec{\tau}$), moment of inertia (I), and angular momentum ($\vec{L}$).

As you may also recall, several of the equations relating various translational quantities have direct parallels in rotational motion. So the rotational equivalent of Newton's Second Law ($\vec{F} = m\vec{a}$) is $\vec{\tau} = I\vec{\alpha}$.[1] And whereas translational momentum is related to mass and velocity by $\vec{p} = m\vec{v}$, you probably learned that angular momentum is related to moment of inertia and angular velocity by $L_z = I\omega$.

When first presenting these relationships, most texts restrict the motion to planar rotation of a single particle to keep things simple. So when you think of the relationship between linear and angular velocity, you may think of something like $v = \omega r$. And if $L_z = mvr$, then $L_z = mr^2\omega$. Taking mr^2 as the moment of inertia (I) of a single particle, this becomes $L_z = I\omega$. But the v and the ω in those equations can't really be velocities, since they're written as scalars rather than vectors, and that z subscript on the angular momentum seems to be trying to tell you something.

It is. It's telling you that you're using an equation for one component of the angular momentum (the z-component in this case), and this pertains to a single particle moving about the origin in the xy plane. So these equations aren't wrong, they just have limited application. Specifically, they apply to cases of planar motion about the z-axis.

The more-general relationship between the vectors that represent velocity, angular velocity, and position is this:

$$\vec{v} = \vec{\omega} \times \vec{r}, \tag{6.1}$$

in which the cross represents the vector cross product described in Chapter 2. And the equations relating angular momentum to linear momentum, linear velocity, and mass are

$$\begin{aligned} \vec{L} &= \vec{r} \times \vec{p} \\ &= \vec{r} \times (m\vec{v}) \\ &= m\vec{r} \times \vec{v}. \end{aligned} \tag{6.2}$$

Before delving more deeply into these equations, you should consider the implications of the (planar-motion) equation that says that the moment of inertia of a single particle is $I_{particle} = mr^2$. One important idea in this equation is that the moment of inertia of a particle depends not only on its mass, but also on the location of that mass – specifically, the distance (r) of the mass from the axis of rotation. Thus the moment of inertia of an extended object made up of many particles must depend not only on the object's mass,

[1] Or, if you prefer the more-general form of Newton's Second Law ($\vec{F} = \frac{d\vec{p}}{dt}$), the analogous rotational relationship is $\vec{\tau} = \frac{d\vec{L}}{dt}$.

but on the distribution of that mass. That's true in the case of general motion as well as planar rotation.

If you think of the rotational analog to the translational equation $\vec{p} = m\vec{v}$, you may be tempted to write an equation such as $\vec{L} = I\vec{\omega}$. But that equation would indicate that the angular momentum $\vec{L}$ must be in the same direction as the angular velocity $\vec{\omega}$, since multiplication by a scalar can change the length but not the direction of a vector (unless the scalar is negative, in which case the direction of the vector is reversed). For general motion, the situation is more complex, as you can see by applying Eq. 6.2 to a single particle circling about the axis shown in Figure 6.1. In this figure, the particle "m" is circling around the z-axis, so the angular velocity ($\vec{\omega}$) points straight up, parallel to the z-axis. In this view, you're looking down the x-axis toward the origin of the coordinate system, which is well below the plane of the particle's path. The particle is initially at the position shown on the left side of the figure, and its velocity vector is coming out of the page. Since the vector angular momentum is given by $\vec{L} = m\vec{r} \times \vec{v}$, you can find the direction of the angular momentum at this initial instant by using your right hand to form the cross product between $\vec{r}$ and $\vec{v}$, as described in Section 2.2. If you do this properly, you should see that $\vec{L}$ initially points up and to the right, as shown by $\vec{L}_{initial}$ in the figure. At a later time, after the particle has completed one-half revolution about the z-axis, its velocity vector is into the page, as shown in the right

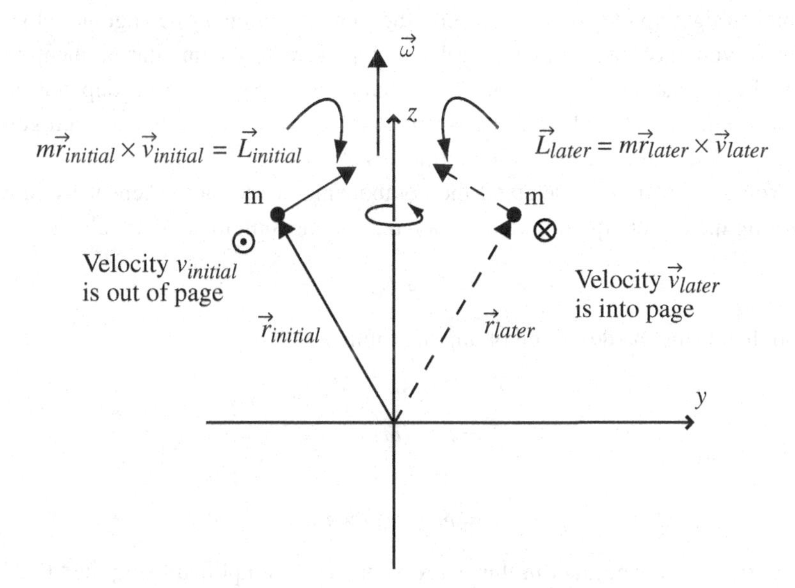

Figure 6.1 Single point mass moving around an axis.

portion of the figure. At that later instant, the cross product between $\vec{r}$ and $\vec{v}$ means that the direction of the angular momentum vector $\vec{L}$ is up and to the left, as shown by $\vec{L}_{later}$.

So not only is the angular-momentum vector $\vec{L}$ not parallel to the angular-velocity vector $\vec{\omega}$, the direction of the $\vec{L}$ is changing as the particle moves around the axis, while the direction of $\vec{\omega}$ remains fixed along the z-axis.

Under these circumstances, you clearly cannot use a scalar value for the moment of inertia to relate the angular momentum to the angular velocity through an equation such as $\vec{L} = I\vec{\omega}$. A scalar moment of inertia simply isn't capable of relating a vector in one direction to a different vector in another direction. But if you've followed the developments of Chapters 4 and 5, you're already familiar with a type of object that is capable of taking in a vector (such as $\vec{\omega}$) and producing another vector (such as $\vec{L}$) that points in a different direction. That object is a tensor. So although you may have initially learned about the moment of inertia as a scalar value in the case of planar motion about the origin, you should now understand why more-general problems require a more-powerful approach, and that involves the representation of inertia as a tensor rather than a scalar.

You may be thinking that simply by adding another particle of equal mass at the same distance on the other side of the z-axis, you could produce an additional bit of angular momentum that would add to the angular momentum of the original mass. In that case, the total angular momentum would indeed point straight up the z-axis, in exactly the same direction as the angular velocity. So you may suspect that the relationship between the angular momentum and the angular velocity (and hence the nature of the inertia tensor) depends on the symmetry of the object. That suspicion is correct, as you'll see when you examine the components of the inertia tensor.

You can begin to understand the components of the inertia tensor by first writing the tensor equation relating angular momentum to angular velocity:

$$\vec{L} = \vec{\vec{I}}\vec{\omega}, \tag{6.3}$$

and then using the definition of angular momentum:

$$\begin{aligned} \vec{L} &= \vec{r} \times \vec{p} \\ &= \vec{r} \times (m\vec{v}) \\ &= m\vec{r} \times \vec{v} \\ &= m\vec{r} \times (\vec{\omega} \times \vec{r}). \end{aligned}$$

The triple vector product in this expression can be simplified using the "BAC minus CAB" rule described in Section 2.4, giving

$$\vec{L} = m[\vec{\omega}(\vec{r} \circ \vec{r}) - \vec{r}(\vec{r} \circ \vec{\omega})].$$

This is a usable expression for the angular momentum of a single particle, and you can modify it for use with multiple masses simply by summing (or for a continuous object by integrating) over all the masses. Thus the expression you'll most often encounter will probably look something like this:

$$\vec{L} = \sum_i m_i[\vec{\omega}(\vec{r}_i \circ \vec{r}_i) - \vec{r}_i(\vec{r}_i \circ \vec{\omega})], \tag{6.4}$$

where the index i denotes each element of mass of the object.

To see the moment of inertia in this expression, first expand the position vector as $\vec{r}_i = x_i\hat{\imath} + y_i\hat{\jmath} + z_i\hat{k}$ and the angular velocity vector as $\vec{\omega} = \omega_x\hat{\imath} + \omega_y\hat{\jmath} + \omega_z\hat{k}$ (note that the angular velocity $\vec{\omega}$ is the same for every mass element in a rigid body, so it's not necessary to write $\vec{\omega}_i$). Thus the expression for angular momentum is

$$\begin{aligned}\vec{L} = \sum_i m_i[&\vec{\omega}(x_i\hat{\imath} + y_i\hat{\jmath} + z_i\hat{k}) \circ (x_i\hat{\imath} + y_i\hat{\jmath} + z_i\hat{k}) \\ &- \vec{r}_i(x_i\hat{\imath} + y_i\hat{\jmath} + z_i\hat{k}) \circ (\omega_x\hat{\imath} + \omega_y\hat{\jmath} + \omega_z\hat{k})],\end{aligned}$$

and performing the dot products gives

$$\vec{L} = \sum_i m_i[\vec{\omega}(x_i^2 + y_i^2 + z_i^2) - \vec{r}_i(x_i\omega_x + y_i\omega_y + z_i\omega_z)].$$

Since the x-component of $\vec{\omega}$ is ω_x and the x-component of $\vec{r}_i$ is x_i, the x-component of the angular momentum can be written

$$\begin{aligned}L_x &= \sum_i m_i[\omega_x(x_i^2 + y_i^2 + z_i^2) - x_i(x_i\omega_x + y_i\omega_y + z_i\omega_z)] \\ &= \sum_i m_i[\omega_x x_i^2 + \omega_x y_i^2 + \omega_x z_i^2 - x_i^2\omega_x - x_i y_i\omega_y - x_i z_i\omega_z] \\ &= \sum_i m_i[\omega_x(y_i^2 + z_i^2) - x_i y_i\omega_y - x_i z_i\omega_z].\end{aligned}$$

The y- and z-components come out as

$$\begin{aligned}L_y &= \sum_i m_i[\omega_y(x_i^2 + z_i^2) - y_i x_i\omega_x - y_i z_i\omega_z], \\ L_z &= \sum_i m_i[\omega_z(x_i^2 + y_i^2) - z_i x_i\omega_x - z_i y_i\omega_y].\end{aligned}$$

These three equations for the components of angular momentum ($\vec{L}$) may be written as a single matrix equation:

$$\begin{pmatrix} L_x \\ L_y \\ L_z \end{pmatrix} = \begin{pmatrix} \sum_i m_i(y_i^2 + z_i^2) & -\sum_i m_i x_i y_i & -\sum_i m_i x_i z_i \\ -\sum_i m_i y_i x_i & \sum_i m_i(x_i^2 + z_i^2) & -\sum_i m_i y_i z_i \\ -\sum_i m_i z_i x_i & -\sum_i m_i z_i y_i & \sum_i m_i(x_i^2 + y_i^2) \end{pmatrix} \begin{pmatrix} \omega_x \\ \omega_y \\ \omega_z \end{pmatrix}. \tag{6.5}$$

The elements of the center matrix represent the components of the inertia tensor ($\vec{\vec{I}}$). Note that the dimensions of each element are mass times distance squared (SI units of kg m^2), just as in the case of scalar moment of inertia.

In some texts, you'll find the elements of the inertia tensor written as something like

$$I_{ab} = m_i(\delta_{ab} r_i^2 - r_a r_b),$$

which are the same elements as shown in Eq. 6.5.

The diagonal elements of the inertia tensor are called "moments of inertia" and the off-diagonal elements are called "products of inertia." To understand the physical meaning of each of these elements, recall that the moment of inertia characterizes an object's tendency to resist angular acceleration. That resistance depends not only on the object's mass, but on the distribution of that mass relative to the axis of rotation.

Each term I_{ab} tells you how much angular momentum in the a-direction is produced by rotation about the b-axis. So $I_{11} = I_{xx}$ tells you how much angular momentum the object produces in the x-direction due to rotation about the x-axis. And $I_{23} = I_{yz}$ tells you how much angular momentum the object produces in the y-direction due to rotation about the z-axis.

How those off-diagonal terms come about is explained below, but you should first take a look at the diagonal terms. In the expression for I_{xx}, for each element of mass (m_i), the element's mass is multiplied by the square of the distance from the x-axis ($y_i^2 + z_i^2$). So this is just the three-dimensional version of the equation you may have learned for planar rotation that says that the moment of inertia of a particle is $I = mr^2$, where r is the particle's distance from the axis of rotation. Looking down the diagonal of the inertia tensor, you see the contribution to the x-component of angular momentum due to rotation about the x-axis, the contribution to the y-component of angular momentum due to rotation about the y-axis, and the contribution to the z-component of angular momentum due to rotation about the z-axis. The bottom line is that distributions of mass that are symmetric about each axis contribute to the diagonal terms of the moment of inertia matrix.

The off-diagonal elements of the inertia tensor are somewhat different. In I_{yz}, for each element of mass (m_i), the element's mass is multiplied by the product of the element's y- and z-coordinates ($y_i z_i$). As explained above, this determines the contribution to the y-component of angular momentum due to rotation about the z-axis. And when does rotation about the z-axis produce a y-component of angular momentum? When there's an asymmetric distribution of mass about the z-axis, for example as shown with the single particle in Figure 6.1. Likewise, the I_{xy} term determines the contribution to the x-component of angular momentum due to rotation about the y-axis. Such contributions come from mass distributions that are asymmetric about the y-axis. Hence distributions of mass that are asymmetric about a given axis contribute to the off-diagonal terms of the moment of inertia matrix.

To see how this works, consider the five point masses on the corners and top of a pyramid as shown in Figure 6.2. To determine the inertia tensor for this configuration of masses, you simply have to plug the mass and coordinates of each of the masses into Equation 6.5. If the mass of each of the five masses is the same and equal to "m" and the height of the pyramid is equal to the length of each of the bottom sides (with a value of $2a$ as shown in Figure 6.2), the I_{xx} term is simply

$$\begin{aligned}
I_{xx} &= m_1(y_1^2 + z_1^2) + m_2(y_2^2 + z_2^2) + m_3(y_3^2 + z_3^2) + m_4(y_4^2 + z_4^2) \\
&\quad + m_5(y_5^2 + z_5^2) \\
&= m_1(a^2 + 0^2) + m_2(a^2 + 0^2) + m_3[(-a)^2 + 0^2] + m_4[(-a)^2 + 0^2] \\
&\quad + m_5(0^2 + (2a)^2) \\
&= 8ma^2,
\end{aligned}$$

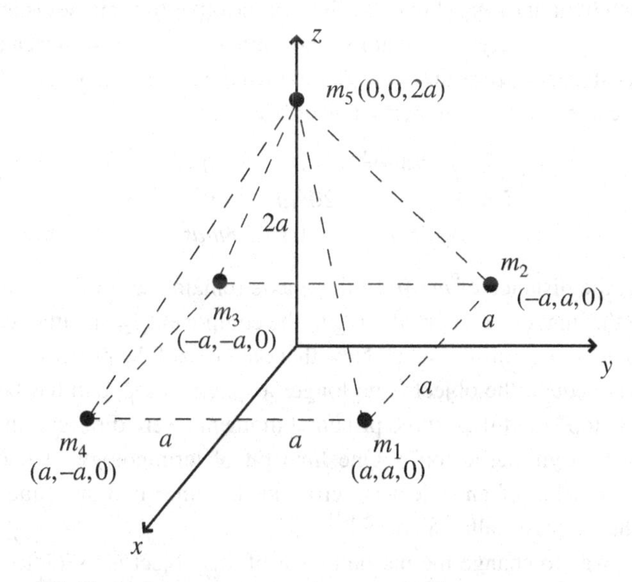

Figure 6.2 Five point masses arrayed as a pyramid.

and you should obtain the same result for the other diagonal elements I_{yy} and I_{zz}. Moving on to the off-diagonal elements, the I_{xy} term is

$$\begin{aligned} I_{xy} &= -m_1x_1y_1 - m_2x_2y_2 - m_3x_3y_3 - m_4x_4y_4 - m_5x_5y_5 \\ &= -m_1(a)(a) - m_2(-a)(a) - m_3(-a)(-a) - m_4(a)(-a) - m_5(0)(0) \\ &= -m(2a^2 - 2a^2) = 0, \end{aligned}$$

which is the same as all other off-diagonal elements. Thus the matrix representing the inertia tensor for the configuration shown in Figure 6.2 is

$$\vec{\vec{I}} = \begin{pmatrix} 8ma^2 & 0 & 0 \\ 0 & 8ma^2 & 0 \\ 0 & 0 & 8ma^2 \end{pmatrix}.$$

There's a great deal of information in the components of this inertia tensor. The fact that the off-diagonal elements are all zero means that the selected x-, y-, and z-axes are "principal axes" for this object and choice of origin, and the moments of inertia are "principal moments" of the object. When an object rotates about one of the principal axes, the angular momentum vector and the angular velocity vector are parallel. This is an indication of the object's symmetry. In this case, the fact that all three principal moments are equal means that this object qualifies as a "spherical top" (in Mechanics, "top" refers to any rigid rotating object). And for a spherical top, any three mutually orthogonal axes are principal axes.

If the height of mass m_5 above the plane of the other four masses is increased to twice its original height (so that its z-coordinate becomes $4a$ instead of $2a$), the greater distance from the x- and y-axes increases the moment of inertia about those axes, so that the inertia tensor becomes

$$\vec{\vec{I}} = \begin{pmatrix} 20ma^2 & 0 & 0 \\ 0 & 20ma^2 & 0 \\ 0 & 0 & 8ma^2 \end{pmatrix}.$$

Of course, the distance of m_5 from the z-axis remains zero irrespective of its height, so this mass is not contributing to the component I_{zz} in either case, and that component remains the same. Now that only two of the principal moments of inertia are equal, the object is no longer a spherical top, and has become a "symmetric top" (and if all three principal moments were different, the object is called an "asymmetric top"). One final bit of terminology: if one of the principal moments of an object is zero and the other two are equal to one another, the object is called a "rotor."

Another way to change the inertia tensor of this object is to fiddle with the masses of the particles. If, for example, you double the mass of m_5 from its original value of m to $2m$, while leaving the other four masses the same, the inertia tensor becomes

$$\vec{\vec{I}} = \begin{pmatrix} 12ma^2 & 0 & 0 \\ 0 & 12ma^2 & 0 \\ 0 & 0 & 8ma^2 \end{pmatrix}.$$

As expected, there's no change in the I_{zz} component since m_5 doesn't contribute to that moment.

Now consider what will happen to the inertia tensor if you rotate the coordinate axes. Remember, the inertia tensor is determined for a given location of the origin and a given orientation of the coordinate axes, so it seems reasonable to expect a change in the components if the coordinate axes are rotated.

To test this, imagine rotating the coordinate axes counter-clockwise about the x-axis, as shown in Figure 6.3. In this figure, you're looking down the x-axis toward the origin, so the y- and z-axes appear tilted (they're labeled y' and z' to distinguish them from the original y- and z-axes). In this case, the rotation angle is approximately 30°. Figure 6.3(a) shows that the axes have rotated while the masses remained in their original positions, while Figure 6.3(b) shows the view you would get if you tilted your head to make the z'-axis vertical and y'-axis horizontal.

What effect might this have on the inertia tensor? To determine that, you'll need to know the coordinates of each of the masses in the new (rotated) coordinate system (that is, you need to know x', y', and z' for each mass). Fortunately, Chapter 4 should have given you some idea of how to do that by using a rotation matrix to convert between the original and rotated coordinates. In this case, that rotation matrix is given by

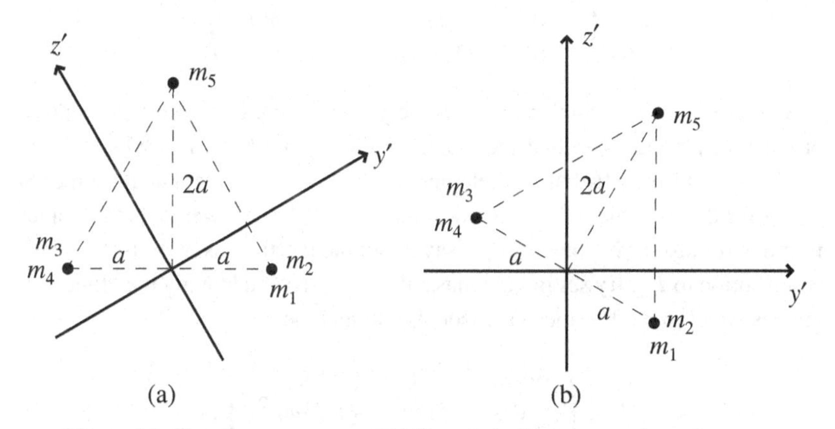

Figure 6.3 Coordinate axes rotated 30° anti-clockwise around x-axis.

$$\begin{pmatrix} x' \\ y' \\ z' \end{pmatrix} = \begin{pmatrix} 1 & 0 & 0 \\ 0 & \cos\theta & \sin\theta \\ 0 & -\sin\theta & \cos\theta \end{pmatrix} \begin{pmatrix} x \\ y \\ z \end{pmatrix}. \tag{6.6}$$

If you go back to the original masses (all five masses equal to mass m) and original height of m_5 (which is $2a$ above the xy plane) and then apply this rotation, you should find the following values for the components of the matrix representing the inertia tensor:

$$\overleftrightarrow{I} = \begin{pmatrix} 8ma^2 & 0 & 0 \\ 0 & 8ma^2 & 0 \\ 0 & 0 & 8ma^2 \end{pmatrix}.$$

If you're suprised to find that there's no change from the original inertia tensor (the one without the rotation), remember that the symmetry of this object makes it a spherical top, which means that any set of three orthogonal axes will be principal axes. So tilting the axes should not have caused any change in the inertia tensor.

That sounds reasonable enough, but if you compare the location of the masses in Figure 6.3 to the single-mass case shown in Figure 6.1, doesn't it also seem reasonable to expect that m_5 will produce a component of angular momentum in the $-y$-direction (as the single mass did in Figure 6.1)?

Yes, it does. And, in fact, mass m_5 does indeed produce a component of angular momentum in the $-y$-direction. To demonstrate that, just set the other four masses to zero and calculate the inertia tensor for m_5 alone (don't forget that the coordinate axes are rotated). You should get

$$\overleftrightarrow{I} = \begin{pmatrix} 4ma^2 & 0 & 0 \\ 0 & 3ma^2 & -1.73ma^2 \\ 0 & -1.73ma^2 & ma^2 \end{pmatrix}.$$

So there it is: I_{yz} (which represents the y-component of angular momentum produced by rotation around the z-axis) is clearly not zero. But why did you get zero for all the off-diagonal elements when you first calculated the inertia tensor for the pyramid with tilted coordinate axes? The answer is that the other four masses also have something to say about the inertia tensor. To isolate their contribution to I_{yz}, try setting the mass of m_5 to zero and leaving the other four masses equal to m. The inertia tensor should then be

$$\overleftrightarrow{I} = \begin{pmatrix} 4ma^2 & 0 & 0 \\ 0 & 5ma^2 & 1.73ma^2 \\ 0 & 1.73ma^2 & 7ma^2 \end{pmatrix}.$$

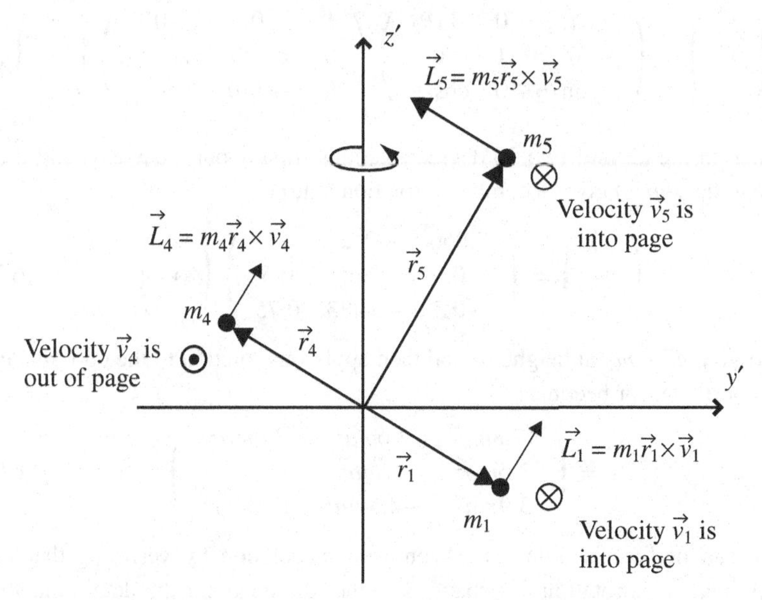

Figure 6.4 Angular momentum vectors for masses in plane of page.

And there's the answer: the other four masses contribute exactly as much angular momentum in the positive y-direction as m_5 contributes to the negative y-direction, as illustrated in Figure 6.4. And remember from Chapter 5 that you can add tensors by adding their components. So when you add the inertia tensor for m_5 to the inertia tensor for the other four masses, you get the (nicely diagonal) inertia tensor for the five-mass pyramid.

To demonstrate the balance between m_5 and the other four masses, you may find it interesting to again move m_5 up the z-axis to twice its original height and then perform the 30 degree rotation of the coordinate axes. In this case, you should find the inertia tensor to be

$$\vec{\vec{I}} = \begin{pmatrix} 20ma^2 & 0 & 0 \\ 0 & 17ma^2 & -5.2ma^2 \\ 0 & -5.2ma^2 & 11ma^2 \end{pmatrix},$$

and clearly the I_{yz} terms from m_5 and the other four masses no longer cancel.

You can determine the inertia tensor for any orientation of the coordinate axes by applying rotations about multiple axes. If you wish, for example, to rotate first about the x-axis by angle θ_1 and then about the y-axis by angle θ_2, you can combine the rotation matrices as

$$\begin{pmatrix} x' \\ y' \\ z' \end{pmatrix} = \begin{pmatrix} \cos\theta_2 & 0 & \sin\theta_2 \\ 0 & 1 & 0 \\ -\sin\theta_2 & 0 & \cos\theta_2 \end{pmatrix} \begin{pmatrix} 1 & 0 & 0 \\ 0 & \cos\theta_1 & \sin\theta_1 \\ 0 & -\sin\theta_1 & \cos\theta_1 \end{pmatrix} \begin{pmatrix} x \\ y \\ z \end{pmatrix}, \tag{6.7}$$

which in the case of two 30 degree rotations (first about the x-axis and then about the y-axis) gives a combined rotation matrix of

$$\begin{pmatrix} x' \\ y' \\ z' \end{pmatrix} = \begin{pmatrix} 0.866 & -0.25 & 4.33 \\ 0 & 0.866 & 0.5 \\ -0.5 & -0.433 & 0.75 \end{pmatrix} \begin{pmatrix} x \\ y \\ z \end{pmatrix}. \tag{6.8}$$

If you leave m_5 at height $4a$ and then apply this rotation to the coordinates, the inertia tensor becomes

$$\vec{\vec{I}} = \begin{pmatrix} 17.8ma^2 & 2.6ma^2 & 3.9ma^2 \\ 2.6ma^2 & 17ma^2 & -4.5ma^2 \\ 3.9ma^2 & -4.5ma^2 & 13.3ma^2 \end{pmatrix}. \tag{6.9}$$

You can perform a quick check on your calculation by verifying that the coordinate-axis rotation has changed neither the trace nor the determinant of the matrix.[2]

Instead of finding the new coordinates of each mass in the rotated system, an alternative approach allows you to find the inertia tensor for rotated coordinates directly. That approach is to apply a "similarity transform" to the original inertia tensor. Here's how that works: the angular momentum is related to the inertia tensor and angular velocity in the original (unrotated) coordinate system as

$$\vec{L} = \vec{\vec{I}}\,\vec{\omega},$$

and you rotate the coordinates by applying a rotation matrix R (which may be the product of several rotation matrices). You can therefore write

$$\vec{L}' = R\vec{L} = R(\vec{\vec{I}}\,\vec{\omega}).$$

And since the product of any matrix and its inverse is just the identity matrix, you can insert the term $R^{-1}R$ in front of $\vec{\omega}$:

$$\begin{aligned} \vec{L}' = R\vec{L} &= R\vec{\vec{I}}\,(R^{-1}R)\vec{\omega} \\ &= (R\vec{\vec{I}}\,R^{-1})R\vec{\omega}. \end{aligned}$$

But $R\vec{\omega}$ is just $\vec{\omega}'$, so

$$\vec{L}' = (R\vec{\vec{I}}\,R^{-1})\vec{\omega}'.$$

[2] The matrix review on the book's website explains how to do these calculations.

Thus the expression $(R\vec{\vec{I}}\,R^{-1})$ relates angular momentum to angular velocity in the rotated coordinate system, which means that this expression is the inertia tensor in that system. So instead of calculating the new coordinates for each mass and plugging them into the equation for the inertia tensor, you can instead simply apply the rotation matrix and its inverse to the matrix representing the inertia tensor directly (but remember that the sequence matters when you're doing matrix multiplication).

Using this approach, the process looks like this:

$$\vec{\vec{I}}\,' = \begin{pmatrix} 0.866 & -0.25 & 4.33 \\ 0 & 0.866 & 0.5 \\ -0.5 & -0.433 & 0.75 \end{pmatrix} \begin{pmatrix} 20ma^2 & 0 & 0 \\ 0 & 20ma^2 & 0 \\ 0 & 0 & 8ma^2 \end{pmatrix}$$

$$\times \begin{pmatrix} 0.866 & -0.25 & 4.33 \\ 0 & 0.866 & 0.5 \\ -0.5 & -0.433 & 0.75 \end{pmatrix}^{-1}$$

$$= \begin{pmatrix} 17.8ma^2 & 2.6ma^2 & 3.9ma^2 \\ 2.6ma^2 & 17ma^2 & -4.5ma^2 \\ 3.9ma^2 & -4.5ma^2 & 13.3ma^2 \end{pmatrix},$$

which is identical to the result obtained by inserting the rotated coordinates into the inertia tensor.

If you've studied matrix algebra, you may be wondering about the possibility of finding the principal axes and principal moments by manipulating the matrix representing the inertia tensor into a diagonal form. That is certainly possible, and you can read about doing that using eigenvectors and eigenvalues on this book's website.

And if you're able by visual inspection to determine the angles of rotation needed to align the axes with the symmetries of the object, you can use the similarity transform approach to diagonalize the inertia matrix. You can see how that works by looking at the problems at the end of this chapter and the on-line solutions.

6.2 The electromagnetic field tensor

One of the defining characteristics of our modern world is the availability of broadband communication channels which allow near-instantaneous transfer of information over great distances without the need for physical connection. The technology used in this communication descends directly from the equations synthesized by Scotsman James Clerk Maxwell in the 1860s, now called "Maxwell's Equations." In view of the impact of electromagnetic telecommunications on our lives, it's not surprising that in 2004 the readers of *Physics World* voted Maxwell's Equations to be the "greatest equations" ever developed.

The four vector equations that have come to be called Maxwell's Equations are Gauss's Law for electric fields, Gauss's Law for magnetic fields, Faraday's Law, and the Ampere–Maxwell Law, each of which may be written in integral or differential form. The integral forms describe the behavior of electric and magnetic fields over surfaces or around paths, while the differential forms apply to specific locations. The differential forms are most relevant to the vector and tensor operations discussed in this book, involving the scalar product, divergence, curl, and partial derivatives discussed in Chapter 2. They're also closely related to the subject of this section, the electromagnetic field-strength tensor.

The differential forms of Maxwell's Equations are usually written as

$$\text{Gauss's Law for electric fields:}\quad \vec{\nabla} \circ \vec{E} = \frac{\rho}{\epsilon_0},$$

$$\text{Gauss's Law for magnetic fields:}\quad \vec{\nabla} \circ \vec{B} = 0,$$

$$\text{Faraday's Law:}\quad \vec{\nabla} \times \vec{E} = -\frac{\partial \vec{B}}{\partial t},$$

$$\text{Ampere–Maxwell Law:}\quad \vec{\nabla} \times \vec{B} = \mu_0 \vec{J} + \mu_0 \epsilon_0 \frac{\partial \vec{E}}{\partial t}.$$

In order to understand the electromagnetic tensor, you may find it helpful to briefly review the meaning of each of these equations.[3]

$$\boxed{\vec{\nabla} \circ \vec{E} = \frac{\rho}{\epsilon_0}}$$

Gauss's Law for electric fields states that the divergence ($\vec{\nabla}\circ$) of the electric field ($\vec{E}$) at any location is proportional to the electric charge density (ρ) at that location. That's because electrostatic field lines begin on positive charge and end on negative charge (hence the field lines tend to diverge away from locations of positive charge and converge toward locations of negative charge).

$$\boxed{\vec{\nabla} \circ \vec{B} = 0}$$

Gauss's Law for magnetic fields tells you that the divergence ($\vec{\nabla}\circ$) of the magnetic field ($\vec{B}$) at any location must be zero. This is true because there is apparently no isolated "magnetic charge" in the universe, so magnetic field lines neither diverge nor converge.

[3] Complete descriptions may be found in any introductory electromagnetics text.

$$\boxed{\vec{\nabla} \times \vec{E} = -\frac{\partial \vec{B}}{\partial t}}$$

Faraday's Law indicates that the curl ($\vec{\nabla}\times$) of the electric field ($\vec{E}$) at any location is equal to the negative of the time rate of change of the magnetic field at that location. That's because a changing magnetic field produces a circulating electric field.

$$\boxed{\vec{\nabla} \times \vec{B} = \mu_0 \vec{J} + \mu_0 \epsilon_0 \frac{\partial \vec{E}}{\partial t}}$$

Ampere's Law, as modified by Maxwell, tells you that the curl ($\vec{\nabla}\times$) of the magnetic field ($\vec{B}$) at any location is proportional to the electric current density ($\vec{J}$) plus the time rate of change of the electric field at that location. This is the case because a circulating magnetic field is produced both by an electric current and by a changing electric field.

Note that Maxwell's Equations relate the spatial behavior of fields to the sources of those fields. Those sources are electric charge (with density ρ) appearing in Gauss's Law for electric fields, electric current (with density $\vec{J}$) appearing in the Ampere–Maxwell Law, changing magnetic field (with time derivative $\frac{\partial \vec{B}}{\partial t}$) appearing in Faraday's Law, and changing electric field (with time derivative $\frac{\partial \vec{E}}{\partial t}$) appearing in the Ampere–Maxwell Law.

One additional equation is needed to fully characterize electromagnetic interactions. That equation is called the "continuity equation," usually written like this:

$$\frac{\partial \rho}{\partial t} = -\vec{\nabla} \circ \vec{J},$$

where ρ is the density of electric charge and $\vec{J}$ is the current density.

The continuity equation tells you that the time rate of change of the density of electric charge ($\frac{\partial \rho}{\partial t}$) equals the negative of the divergence of the electric current density ($\vec{\nabla} \circ \vec{J}$). That's because negative divergence means convergence, and if the convergence of the current density $\vec{J}$ is positive at a point, then more positive charge must be arriving at that location than is being carried away. If that's happening, then the density of positive charge at that point must increase (meaning that $\frac{\partial \rho}{\partial t}$ will be positive in this case).

As valuable as Maxwell's Equations are individually, the real power of these equations is realized by combining them together to produce the wave equation. Taking the curl of both sides of Faraday's Law and inserting the curl of $\vec{B}$ from the Ampere–Maxwell Law results in the equation

$$\nabla^2 \vec{E} = \mu_0 \epsilon_0 \frac{\partial^2 \vec{E}}{\partial t^2}, \tag{6.10}$$

where $\nabla^2() = \vec{\nabla} \circ \vec{\nabla}()$ is the vector form of the Laplacian operator.[4] This equation applies to regions in which the charge density (ρ) and the current density ($\vec{J}$) are both zero.

You can find a similar equation for the magnetic field by taking the curl of both sides of the Ampere–Maxwell Law and then inserting the curl of $\vec{E}$ from Faraday's Law. This gives

$$\nabla^2 \vec{B} = \mu_0 \epsilon_0 \frac{\partial^2 \vec{B}}{\partial t^2}. \tag{6.11}$$

It's instructive to compare Eqs. 6.10 and 6.11 to the general equation for a propagating wave:

$$\nabla^2 \vec{A} = \frac{1}{v^2} \frac{\partial^2 \vec{A}}{\partial t^2}, \tag{6.12}$$

where v is the speed of propagation of the wave. Note the $1/v^2$ term, which leads to the conclusion that the velocity of an electromagnetic wave depends only on the electric permittivity (ϵ_0) and magnetic permeability (μ_0) of free space (specifically, $\mu_0 \epsilon_0 = 1/v^2$, or $v = 1/\sqrt{\mu_o \epsilon_0} = 3 \times 10^8$ m/s). Most importantly, that velocity is completely independent of the motion of the observer. It was this feature of electromagnetic waves that put Albert Einstein onto the path that eventually led to the Theory of Special Relativity.

To arrive at the Theory of Special Relativity, Einstein held fast to two postulates. Those postulates are:

1) The laws of physics must be the same in all inertial (that is, non-accelerating) frames of reference.
2) The speed of light in a vacuum is constant and does not depend on the motion of the source or observer.

Steadfast faithfulness to these postulates even in the face of counter-intuitive conclusions allowed Einstein to see that distances in space and intervals of time are not absolute but depend on the relative motion of the observer. Additionally, space and time are not separate but are linked together into four-dimensional spacetime, and it is the four-dimensional spacetime interval that is invariant across all inertial reference frames.

To understand Einstein's approach, consider the two Cartesian reference frames shown in Figure 6.5. As indicated by the arrow in the figure, the primed reference frame is moving with velocity $\vec{v}$ in the positive x-direction. Using the traditional Galilean approach, the coordinate (x, y, and z) and time (t) values

[4] If you'd like to see the details of the derivation of the electromagnetic-wave equation, you'll find them in the on-line solutions to the problems at the end of this chapter.

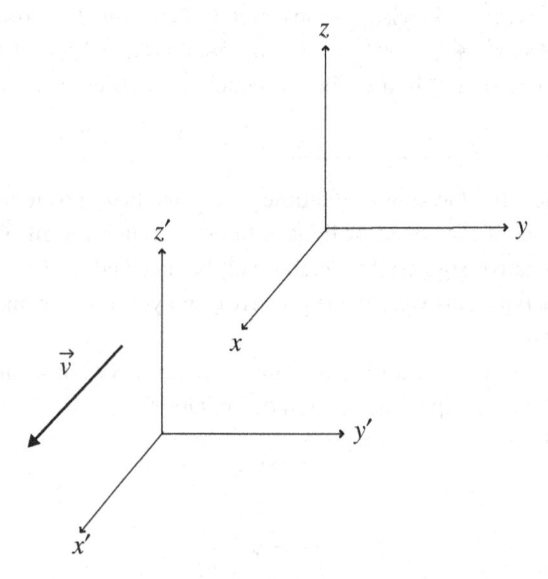

Figure 6.5 Primed reference frame moving along x-axis with velocity $\vec{v}$.

for a point measured in both the unprimed and primed coordinate systems are related by these equations:

$$
\begin{aligned}
t' &= t, \\
x' &= x - vt, \\
y' &= y, \\
z' &= z,
\end{aligned}
$$

since the primed frame is moving only in the x-direction.[5]

Einstein realized that the second postulate of Special Relativity (the constancy of the speed of light) is inconsistent with the Galilean transform shown above, and that consistent results are obtained only when a different transform is used between the unprimed and primed coordinate systems. That transform must hold the space–time interval invariant across inertial reference frames. But what exactly is the space–time interval (that is, how should you combine the space terms and the time term)?

The answer to that question can be understood by imagining a pulse of light radiating spherically outward from a certain location. Calling the speed of light c, an observer in the unprimed coordinate system will find the square of the distance covered by a wavefront of the light wave in time t to be

[5] These equations assume that the origins of the two coordinate systems coincide at time $t = 0$.

$x^2 + y^2 + z^2 = ct^2$. Likewise, an observer in the primed coordinate system will write this as $x'^2 + y'^2 + z'^2 = ct'^2$. But by the second postulate of special relativity, the speed of light must be the same for all observers. So

$$ct^2 - x^2 - y^2 - z^2 = ct'^2 - x'^2 - y'^2 - z'^2,$$

which indicates that the sign of the time term must be opposite to the sign of the spatial terms if the speed of light is to be the same for all observers. Of course, the negative sign could equally well be attached to the time term (as long as the spatial terms were made positive), and you'll find some texts using that convention.

The combination of one time and three spatial coordinates into a single "four-vector" is best expressed using index notation:

$$x_0 = ct,$$
$$x_1 = x,$$
$$x_2 = y,$$
$$x_3 = z,$$

in which the speed of light (c) is used in the time term to ensure that all four coordinates have dimensions of length.

Using this notation, the space–time interval (ds) can be written as

$$(ds)^2 = (dx^0)^2 - (dx^1)^2 - (dx^2)^2 - (dx^3)^2.$$

This interval is the space–time equivalent of distance ($ds^2 = dx^2 + dy^2 + dz^2$) in three-dimensional space.

Transformations that preserve the invariance of the space–time interval across inertial reference frames are called "Lorentz transforms" after the Dutch physicist Hendrik Lorentz. For motion in $+x$-direction with speed v, the Lorentz transformation is

$$x'_0 = \gamma(x_0 - \beta x_1),$$
$$x'_1 = \gamma(x_1 - \beta x_0),$$
$$x'_2 = x_2,$$
$$x'_3 = x_3,$$

where

$$\beta = \frac{|v|}{c},$$

and

$$\gamma = \frac{1}{\sqrt{1 - \frac{v^2}{c^2}}} = \frac{1}{\sqrt{1 - \beta^2}}.$$

This form of the space–time interval can be written using the metric tensor $g_{\alpha\beta}$:

$$(ds)^2 = g_{\alpha\beta} dx^\alpha dx^\beta,$$

in which the tensor $g_{\alpha\beta}$ corresponds to the Minkowski metric for flat space-time. In matrix form, that metric is

$$\overleftrightarrow{g} = \begin{pmatrix} 1 & 0 & 0 & 0 \\ 0 & -1 & 0 & 0 \\ 0 & 0 & -1 & 0 \\ 0 & 0 & 0 & -1 \end{pmatrix}.$$

As you may recall if you've studied modern physics, the invariance of the space–time interval under Lorentz tranformation leads to several interesting results for observers in different inertial reference frames. Those results include:

(1) Length contraction: An observer in a given reference frame measures lengths in a moving reference frame to be contracted along the direction of motion.
(2) Time dilation: An observer in a given reference frame measures time in a moving reference frame to run more slowly.
(3) Relativity of simultaneity: An observer in a given reference frame will not agree with an observer in a moving reference frame as to whether two events are simultaneous.

Writing physical laws in a form that clearly fits within the framework of Special Relativity has several benefits: such "manifestly covariant" laws have the same form in all inertial reference frames, and the quantities involved transform between reference frames in predictable ways. Any covariant theory of electromagnetism must incorporate the experimental fact that quantity of charge is a scalar (invariant between reference frames), and that Maxwell's Equations and the Lorentz force law are true in all inertial reference frames. This requires a tensor version of the electromagnetic field equations and a four-vector version of the Lorentz force law, which can be accomplished by expressing the electric charge density ρ and current density $\vec{J}$ as a four-vector called the "four-current":

$$\vec{J} = (c\rho, J_x, J_y, J_z).$$

With the four-current in hand, a tensor version of Maxwell's Equations can be achieved by combining the components of the electric and magnetic field into an "electromagnetic field tensor." The matrix representing the contravariant version of this tensor is[6]

$$F^{\alpha\beta} = \begin{pmatrix} 0 & -E_x/c & -E_y/c & -E_z/c \\ E_x/c & 0 & -B_z & B_y \\ E_y/c & B_z & 0 & -B_x \\ E_z/c & -B_y & B_x & 0 \end{pmatrix}. \tag{6.13}$$

The covariant version of this tensor can be found by lowering the indices using the metric tensor. The result is

$$F_{\alpha\beta} = \begin{pmatrix} 0 & E_x/c & E_y/c & E_z/c \\ -E_x/c & 0 & -B_z & B_y \\ -E_y/c & B_z & 0 & -B_x \\ -E_z/c & -B_y & B_x & 0 \end{pmatrix}. \tag{6.14}$$

Another useful tensor is the dual contravariant electromagnetic field tensor

$$\mathfrak{F}^{\alpha\beta} = \begin{pmatrix} 0 & -B_x & -B_y & -B_z \\ B_x & 0 & E_z/c & -E_y/c \\ B_y & -E_z/c & 0 & E_x/c \\ B_z & E_y/c & -E_x/c & 0 \end{pmatrix}. \tag{6.15}$$

One benefit of these tensor expressions is that all of Maxwell's Equations may now be expressed using just two tensor equations. Those two equations are:

$$\frac{\partial F^{\alpha\beta}}{\partial x^{\alpha}} = \mu_0 J^{\beta}, \tag{6.16}$$

and

$$\frac{\partial \mathfrak{F}^{\alpha\beta}}{\partial x^{\alpha}} = 0. \tag{6.17}$$

Where are Maxwell's Equations in these expressions? Well, to find Gauss's Law for electric fields, take $\beta = 0$ in Eq. 6.16:

$$\frac{\partial F^{\alpha 0}}{\partial x^{\alpha}} = \mu_0 J^0.$$

[6] You should be aware that there are almost as many versions of this matrix as there are authors; this book's website has an explanation of the reasons for the differences between the versions found in several popular texts.

Inserting the values from the electromagnetic field-strength tensor of Eq. 6.13 and summing over the dummy index α gives

$$\frac{\partial(0)}{\partial(ct)} + \frac{\partial(E_x/c)}{\partial x} + \frac{\partial(E_y/c)}{\partial y} + \frac{\partial(E_z/c)}{\partial z} = \mu_0(c\rho).$$

Thus

$$\frac{\partial(E_x)}{\partial x} + \frac{\partial(E_y)}{\partial y} + \frac{\partial(E_z)}{\partial z} = \mu_0(c^2\rho),$$

and, since $c^2 = 1/(\epsilon_0\mu_0)$,

$$\frac{\partial(E_x)}{\partial x} + \frac{\partial(E_y)}{\partial y} + \frac{\partial(E_z)}{\partial z} = \frac{\mu_0}{\epsilon_0\mu_0}\rho,$$

or

$$\vec{\nabla} \circ \vec{E} = \frac{\rho}{\epsilon_0},$$

which is Gauss's Law for electric fields.

To get the Ampere–Maxwell Law, look at the equations that result from setting β equal to 1, 2, and 3 in Eq. 6.16:

$$\frac{\partial F^{\alpha 1}}{\partial x^\alpha} = \mu_0 J^1,$$
$$\frac{\partial F^{\alpha 2}}{\partial x^\alpha} = \mu_0 J^2,$$
$$\frac{\partial F^{\alpha 3}}{\partial x^\alpha} = \mu_0 J^3.$$

As above, just insert the values from the electromagnetic field-strength tensor of Eq. 6.13 and sum over the dummy index α:

$$\frac{\partial(-E_x/c)}{\partial(ct)} + \frac{\partial(0)}{\partial x} + \frac{\partial(B_z)}{\partial y} + \frac{\partial(-B_y)}{\partial z} = \mu_0(J_x),$$
$$\frac{\partial(-E_y/c)}{\partial(ct)} + \frac{\partial(-B_z)}{\partial x} + \frac{\partial(0)}{\partial y} + \frac{\partial(B_x)}{\partial z} = \mu_0(J_y),$$
$$\frac{\partial(-E_z/c)}{\partial(ct)} + \frac{\partial(B_y)}{\partial x} + \frac{\partial(-B_x)}{\partial y} + \frac{\partial(0)}{\partial z} = \mu_0(J_z).$$

Hence

$$\frac{\partial(B_z)}{\partial y} - \frac{\partial(B_y)}{\partial z} = \mu_0(J_x) + \frac{1}{c^2}\frac{\partial(E_x)}{\partial t},$$
$$\frac{\partial(B_x)}{\partial z} - \frac{\partial(B_z)}{\partial x} = \mu_0(J_y) + \frac{1}{c^2}\frac{\partial(E_y)}{\partial t},$$
$$\frac{\partial(B_y)}{\partial x} - \frac{\partial(B_x)}{\partial y} = \mu_0(J_z) + \frac{1}{c^2}\frac{\partial(E_z)}{\partial t}.$$

Recognizing the partial derivatives of the magnetic field as the components of the curl of $\vec{B}$, this is

$$\vec{\nabla} \times \vec{B} = \mu_0 \vec{J} + \mu_0 \epsilon_0 \frac{\partial \vec{E}}{\partial t},$$

the Ampere–Maxwell Law.

The other two Maxwell Equations (Gauss's Law for magnetic fields and Faraday's Law) may be obtained in a similar fashion using the dual electromagnetic field-strength tensor (Eq. 6.15). For example, to find Gauss's Law for magnetic fields, take $\beta = 0$ in Eq. 6.17:

$$\frac{\partial \mathfrak{F}^{\alpha 0}}{\partial x^{\alpha}} = 0.$$

Inserting the values from the dual electromagnetic field-strength tensor of Eq. 6.15 and summing over the dummy index α gives

$$\frac{\partial(0)}{\partial(ct)} + \frac{\partial(B_x)}{\partial x} + \frac{\partial(B_y)}{\partial y} + \frac{\partial(B_z)}{\partial z} = 0,$$

which is

$$\vec{\nabla} \circ \vec{B} = 0,$$

Gauss's Law for magnetic fields.

And to get Faraday's Law, look at the equations that result from setting β equal to 1, 2, and 3 in Eq. 6.17:

$$\frac{\partial \mathfrak{F}^{\alpha 1}}{\partial x^{\alpha}} = 0,$$
$$\frac{\partial \mathfrak{F}^{\alpha 2}}{\partial x^{\alpha}} = 0,$$
$$\frac{\partial \mathfrak{F}^{\alpha 3}}{\partial x^{\alpha}} = 0.$$

As before, just insert the values from the dual electromagnetic field-strength tensor of Eq. 6.15 and sum over the dummy index α:

$$\frac{\partial(-B_x)}{\partial(ct)} + \frac{\partial(0)}{\partial x} + \frac{\partial(-E_z/c)}{\partial y} + \frac{\partial(E_y/c)}{\partial z} = 0,$$
$$\frac{\partial(-B_y)}{\partial(ct)} + \frac{\partial(E_z/c)}{\partial x} + \frac{\partial(0)}{\partial y} + \frac{\partial(-E_x/c)}{\partial z} = 0,$$
$$\frac{\partial(-B_z)}{\partial(ct)} + \frac{\partial(-E_y/c)}{\partial x} + \frac{\partial(E_x/c)}{\partial y} + \frac{\partial(0)}{\partial z} = 0.$$

So

$$\frac{\partial(E_y)}{\partial z} - \frac{\partial(E_z)}{\partial y} = \frac{\partial(B_x)}{\partial t},$$
$$\frac{\partial(E_z)}{\partial x} - \frac{\partial(E_x)}{\partial z} = \frac{\partial(B_y)}{\partial t},$$
$$\frac{\partial(E_x)}{\partial y} - \frac{\partial(E_y)}{\partial x} = \frac{\partial(B_z)}{\partial t}.$$

Recognizing the partial derivatives of the electric field as the components of the curl of $\vec{E}$, this is Faraday's Law:

$$\vec{\nabla} \times \vec{E} = -\frac{\partial \vec{B}}{\partial t}.$$

So the use of tensors allows you to write Maxwell's Equations in a simpler form. But the real power of tensors is to help you understand the behavior of electric and magnetic fields when viewed from different reference frames. Specifically, by transforming to a moving reference frame, it becomes clear that electric and magnetic fields depend on the state of motion of the observer.

To see how that comes about, imagine an observer in a reference frame moving along the positive x-axis at a constant speed v. You can investigate the behavior of electric and magnetic fields as seen by this observer by transforming the electromagnetic field tensor to the observer's reference frame.

Recall the Lorentz transform matrix for motion along the x-axis with speed v:

$$A = \begin{pmatrix} \gamma & -\gamma\beta & 0 & 0 \\ -\gamma\beta & \gamma & 0 & 0 \\ 0 & 0 & 1 & 0 \\ 0 & 0 & 0 & 1 \end{pmatrix}. \tag{6.18}$$

So to transform to the primed coordinate system, use

$$\overset{\Rightarrow}{F'} = A\overset{\Rightarrow}{F}A^T,$$

which is

$$\overset{\Rightarrow}{F'} = \begin{pmatrix} \gamma & -\gamma\beta & 0 & 0 \\ -\gamma\beta & \gamma & 0 & 0 \\ 0 & 0 & 1 & 0 \\ 0 & 0 & 0 & 1 \end{pmatrix} \begin{pmatrix} 0 & -E_x/c & -E_y/c & -E_z/c \\ E_x/c & 0 & -B_z & B_y \\ E_y/c & B_z & 0 & -B_x \\ E_z/c & -B_y & B_x & 0 \end{pmatrix}$$
$$\times \begin{pmatrix} \gamma & -\gamma\beta & 0 & 0 \\ -\gamma\beta & \gamma & 0 & 0 \\ 0 & 0 & 1 & 0 \\ 0 & 0 & 0 & 1 \end{pmatrix}.$$

Multiplying the center matrix by the right matrix gives

$$\begin{pmatrix} (-E_x/c)(-\gamma\beta) & (-E_x/c)(\gamma) & -E_y/c & -E_z/c \\ (E_x/c)(\gamma) & (E_x/c)(-\gamma\beta) & -B_z & B_y \\ (E_y/c)(\gamma)+(B_z)(-\gamma\beta) & (E_y/c)(-\gamma\beta)+(B_z)(\gamma) & 0 & -B_x \\ (E_z/c)(\gamma)+(-B_y)(-\gamma\beta) & (E_z/c)(-\gamma\beta)+(B_y)(-\gamma) & B_x & 0 \end{pmatrix},$$

which, when multiplied by the left array, gives

$$\begin{pmatrix} (E_x/c)\gamma^2\beta-(E_x/c)\gamma^2\beta & -(E_x/c)\gamma^2+(E_x/c)\gamma^2\beta^2 \\ (E_x/c)\gamma^2-(E_x/c)\gamma^2\beta^2 & 0 \\ (E_y/c)\gamma-(B_z)\gamma\beta & -(E_y/c)\gamma\beta+(B_z)\gamma \\ (E_z/c)\gamma+(B_y)\gamma\beta & -(E_z/c)\gamma\beta-(B_y)\gamma \end{pmatrix.}$$

$$\begin{matrix} -(E_y/c)\gamma+(B_z)\gamma\beta & -(E_z/c)\gamma-(B_y)\gamma\beta \\ (E_y/c)\gamma\beta-(B_z)\gamma & (E_z/c)\gamma\beta+(B_y)\gamma \\ 0 & -B_x \\ B_x & 0 \end{matrix}\Bigg).$$

Thus

$$\vec{\vec{F}}' = \begin{pmatrix} 0 & -E_x/c \\ E_x/c & 0 \\ \gamma(E_y/c-\beta B_z) & \gamma(B_z-\beta E_y/c) \\ \gamma(E_z/c+\beta B_y) & -\gamma(B_y+\beta E_z/c) \end{matrix}$$

$$\begin{matrix} \gamma(E_y/c-\beta B_z) & -\gamma(E_z/c+\beta B_y) \\ -\gamma(B_z-\beta E_y/c) & \gamma(B_y+\beta E_z/c) \\ 0 & -B_x \\ B_x & 0 \end{matrix}\Bigg).$$

Comparing this to Eq. 6.13, the components of the electric field in the new (primed) coordinate system can be related to the components of the electric field in the original (unprimed) coordinate system by

$$\begin{aligned} E'_x &= E_x, \\ E'_y &= c\gamma(E_y/c-\beta B_z), \\ E'_z &= c\gamma(E_z/c+\beta B_y), \end{aligned} \tag{6.19}$$

and the magnetic field components in the new (primed) system are

$$\begin{aligned} B'_x &= B_x, \\ B'_y &= \gamma(B_y+\beta E_z/c), \\ B'_z &= \gamma(B_z-\beta E_y/c). \end{aligned} \tag{6.20}$$

This is a profound result, since it indicates that the existence of electric and magnetic fields depends on the motion of the observer.

To understand the implications of these results, consider the case in which $E_x = E_y = E_z = 0$ but one or more components of $\vec{B}$ are non-zero (this occurs, for example, when a long, straight wire carries a steady electric current). This means that an observer in the unprimed coordinate system sees a magnetic field but no electric field. However, transforming to the primed coordinate system, Eqs. 6.19 and 6.20 tell you that an observer in the primed coordinate system sees both electric and magnetic fields (since in this case $E'_y = -c\gamma\beta B_z$ and $E'_z = c\gamma\beta B_y$). So does the magnetic field exist or not? The answer depends on the motion of the observer.

Now consider a case in which $B_x = B_y = B_z = 0$ but one or more components of $\vec{E}$ are non-zero in the unprimed system (for example, an electric charge at rest in the unprimed system). For this case, an observer in the primed system does see a magnetic field with components $B'_y = \gamma\beta E_z/c$ and $B'_z = -\gamma\beta E_y/c$ (this makes sense, since the observer in the primed system sees a moving electric charge, which is an electric current, and electric currents produce magnetic fields). Cases such as these explain the reasoning behind the statement that electric and magnetic fields "have no independent existence."

The problems at the end of this chapter will give you an idea of the relative magnitudes of fields seen by an observer at rest and a second observer moving at a significant fraction of the speed of light.

6.3 The Riemann curvature tensor

In the decade after publishing his Theory of Special Relativity in 1905, Albert Einstein turned his attention to what he called a "deficiency" in classical mechanics: the lack of an explanation for the precise equality of inertial and gravitational mass. An object's inertial mass determines its resistance to acceleration, and its gravitational mass determines its response to a gravitational field. The equality of these differently defined masses cannot be explained by classical mechanics, and Einstein's scientific instincts told him that the resolution of this deficiency could be achieved by "an extension of the principle of relativity to spaces of reference which are not in uniform motion relative to one another."[7] He applied the word "General" to this extension of his theory of relativity because this new theory would not be restricted to the non-accelerating reference frames of Special Relativity.

[7] A. Einstein, *The Meaning of Relativity*.

Early in his work on the General Theory, Einstein constructed a Gedanken-experiment (that is, a mental exercise) in which he imagined a group of objects with different mass far away from the Earth and from all other masses – you can think of this as a bunch of rocks far out in space. The behavior of these objects is observed from two reference systems, one of which is called system K and is "inertial" or non-accelerating with respect to the rocks. The other system, called system K′, is in uniform acceleration with respect to the first. For an observer in the K′ system, the objects all accelerate in the same direction (opposite to the direction of the acceleration of the K′ system) and at the same rate (equal to the rate of acceleration of the K′ system). Seeing all objects accelerating in the same direction and at the same rate, that observer would be entirely justified in concluding that the acceleration of the objects is produced by an external gravitational field and that the K′ system is at rest. Einstein realized that both the K and the K′ systems are valid frames of reference, and he termed the complete equivalence of such systems the "principle of equivalence."

Einstein's next step was to overlay the z'-axis of K′ system with the z-axis of the K system and then to allow the K′ system to rotate about the z'-axis with uniform angular speed (recall that a rotating object experiences centripetal acceleration, so rotation makes K′ an accelerated system). If system K′ were not rotating, the size of objects and rate of time flow measured in the K and K′ systems would be the same. But when system K′ is rotating, objects at rest in K′ will be moving when measured in the K system and will therefore experience length contraction and time dilation, and the amount of contraction and dilation will depend on the location of the objects (since objects farther from the rotation axis will have higher velocity). Since the principle of equivalence demands that an accelerated system and a system at rest in a gravitational field are equivalent, Einstein was forced to conclude that length contraction and time dilation could also be produced by gravity, or as he put it "the gravitational field influences and even determines the metrical laws of the space–time continuum."

Those metrical laws are expressed using tensors, so the General Theory of Relativity relies on tensor formulation of physical laws and on concepts described in earlier chapters, such as the metric tensor, Christoffel symbols, and covariant derivatives. The most important tensor in General Relativity is the Riemann curvature tensor, sometimes called the Riemann–Christoffel tensor after the nineteenth-century German mathematicians Bernhard Riemann and Elwin Bruno Christoffel. The importance of this tensor stems from the fact that non-zero components are the hallmark of curvature; the vanishing of the Riemann tensor is both a necessary and a sufficient condition for Euclidean (flat) space.

Most texts use one of two ways to derive the Riemann curvature tensor: parallel transport or the commutator of the covariant derivative. To understand the parallel-transport approach, you should first understand that "parallel transport" refers to a method of moving a vector around a space while keeping the length and direction of the vector the same. In Cartesian flat space, making sure the vector's magnitude and direction don't change is straightforward – just move the vector around without allowing the x-, y-, or -z components to change. If the components don't change, then the length and the direction of the vector don't change, and this satisfies the requirements of parallel transport.

In curved spaced, the situation is more complex. For one thing, "pointing in the same direction" becomes more difficult to define. Consider the two-dimensional space that is the surface of the Earth (and pretend for the moment that it's perfectly smooth). Imagine a vector that is initially at the equator (say a bit north of Quito, Ecuador) and is pointing due north, directly along the meridian line. Now imagine transporting that vector toward the north pole, all the while making sure it remains pointed exactly along the meridian line. Remember, the entire space is the surface of the Earth, so the vector must remain tangent to the surface (that is, locally horizontal) as you move it. If you continue moving your vector along the meridian line and pass over the North Pole and then "down" the other side of the Earth, you will eventually reach the equator again somewhere near the middle of Indonesia. Your vector will still be pointing along the meridian, but now it will be pointing south. So although you've kept your vector pointing "in the same direction" (that is, along the meridian) over the entire trip, it's gone from pointing north to pointing south.

Now imagine making another trip, also starting with a north-pointing vector at the equator near Quito, but this time moving along the equator instead of over the North Pole. Once again, as you move you make sure that your vector continues to point north (along the local meridian). After a long journey, you arrive in the middle of Indonesia, but this time you find that your vector is pointing north. Hence the direction of the vector at the end of the journey depends on the path taken, even though you used parallel transport in each case. And whenever the result of parallel transport is a change in the direction of a vector, you can be sure you're dealing with a curved space.

This raises a larger issue: it's not possible to add, subtract, multiply, or in any way compare vectors at different locations – you have to transport one of the vectors to the location of the other before you can perform such operations. That's no problem in flat space, because you can parallel-transport a vector to any other location simply by keeping its coefficients constant (ensuring that the vector's length is constant and that it remains pointed in the same direction). But while "pointed in the same direction" is easily defined at different locations in flat space, you've just seen that this phrase is problematic in curved space. Hence a more-general definition of parallel transport is required.

In that definition, "parallel transport" is defined as transport for which the covariant derivative is zero. Remember that the covariant derivative is the combination of two terms, the first of which is the usual partial derivative, and the second of which involves a Christoffel symbol. As described in Section 5.7 in Chapter 5, the purpose of that second term is to account for changes in the basis vectors. Holding the covariant derivative at zero while transporting a vector around a small loop is one way to derive the Riemann tensor.[8]

The Riemann curvature tensor falls naturally out of the commutator of the covariant derivative of a vector. In this usage, "commutator" refers to the difference that results from performing two operations first in one order and then in the reverse order. So if one operator is denoted by A and another operator by B, the commutator is defined as [AB] = AB−BA. Thus if the sequence of the two operations has no impact on the result, the commutator has a value of zero.

To get to the Riemann tensor, the operation of choice is covariant differentiation. That's because in a flat space the order of covariant differentiation makes no difference, so the commutator must yield zero. Any non-zero result of applying the commutator to covariant differentiation can therefore be attributed to the curvature of the space.

To begin this process, take the covariant derivative of vector V_α first with respect to x^β:

$$V_{\alpha;\beta} = \frac{\partial V_\alpha}{\partial x^\beta} - \Gamma^\sigma_{\alpha\beta} V_\sigma. \tag{6.21}$$

Now call this result $V_{\alpha\beta}$ and take another covariant derivative (this time with respect to x^γ):

$$V_{\alpha\beta;\gamma} = \frac{\partial V_{\alpha\beta}}{\partial x^\gamma} - \Gamma^\tau_{\alpha\gamma} V_{\tau\beta} - \Gamma^\eta_{\beta\gamma} V_{\alpha\eta}. \tag{6.22}$$

Substituting the expression from Eq. 6.21 into this equation gives

$$\begin{aligned} V_{\alpha\beta;\gamma} = & \frac{\partial^2 V_\alpha}{\partial x^\gamma \partial x^\beta} - \frac{\partial \Gamma^\sigma_{\alpha\beta}}{\partial x^\gamma} V_\sigma - \Gamma^\sigma_{\alpha\beta} \frac{\partial V_\sigma}{\partial x^\gamma} \\ & - \Gamma^\tau_{\alpha\gamma} \left(\frac{\partial V_\tau}{\partial x^\beta} - \Gamma^\sigma_{\tau\beta} V_\sigma \right) \\ & - \Gamma^\eta_{\beta\gamma} \left(\frac{\partial V_\alpha}{\partial x^\eta} - \Gamma^\sigma_{\alpha\eta} V_\sigma \right). \end{aligned} \tag{6.23}$$

[8] You can find the details in Schutz, *A First Course in General Relativity*, Cambridge University Press, 2009.

It's not easy to see the physical significance in this expression, but remember how you got here: first by finding the incremental change in V_α as you take a small step in the x^β-direction, and then finding the change in that quantity as you take a small step in the x^γ-direction. And now you're going to compare the result of these two operations with the result you get when you take the steps in reverse order – from the same starting point, you'll first find the incremental change in V_α as you take a small step in the x^γ-direction, after which you'll find the change in that quantity as you take a small step in the x^β-direction.

To take the covariant derivatives in the opposite order, differentiate first with respect to x^γ:

$$V_{\alpha;\gamma} = \frac{\partial V_\alpha}{\partial x^\gamma} - \Gamma^\sigma_{\alpha\gamma} V_\sigma. \tag{6.24}$$

Call this result $V_{\alpha\gamma}$ and take another covariant derivative (this time with respect to x^β):

$$V_{\alpha\gamma;\beta} = \frac{\partial V_{\alpha\gamma}}{\partial x^\beta} - \Gamma^\tau_{\alpha\beta} V_{\tau\gamma} - \Gamma^\eta_{\gamma\beta} V_{\alpha\eta}. \tag{6.25}$$

As before, you can substitute the expression from Eq. 6.24 into this equation to get

$$\begin{aligned} V_{\alpha\gamma;\beta} = {} & \frac{\partial^2 V_\alpha}{\partial x^\beta \partial x^\gamma} - \frac{\partial \Gamma^\sigma_{\alpha\gamma}}{\partial x^\beta} V_\sigma - \Gamma^\sigma_{\alpha\gamma} \frac{\partial V_\sigma}{\partial x^\beta} \\ & - \Gamma^\tau_{\alpha\beta} \left(\frac{\partial V_\tau}{\partial x^\gamma} - \Gamma^\sigma_{\tau\gamma} V_\sigma \right) \\ & - \Gamma^\eta_{\gamma\beta} \left(\frac{\partial V_\alpha}{\partial x^\eta} - \Gamma^\sigma_{\alpha\eta} V_\sigma \right). \end{aligned} \tag{6.26}$$

In flat space, the order of covariant differentiation should make no difference, so Eq. 6.26 should be identical to Eq. 6.23. Any differences between these equations can therefore be attributed to the curvature of the space. Examining these two equations term by term, the first terms are equal:

$$\frac{\partial^2 V_\alpha}{\partial x^\gamma \partial x^\beta} = \frac{\partial^2 V_\alpha}{\partial x^\beta \partial x^\gamma},$$

(these terms are equal because the order of normal partial derivatives does not matter). Hence these terms cancel in the commutator. Now comparing the second terms,

$$-\frac{\partial \Gamma^\sigma_{\alpha\beta}}{\partial x^\gamma} V_\sigma \neq -\frac{\partial \Gamma^\sigma_{\alpha\gamma}}{\partial x^\beta} V_\sigma,$$

so these terms do not cancel one another. Comparing the third term of Eq. 6.23 to the fourth term of Eq. 6.26, they're found to be equal:

$$-\Gamma^{\sigma}_{\alpha\beta}\frac{\partial V_{\sigma}}{\partial x^{\gamma}} = -\Gamma^{\tau}_{\alpha\beta}\frac{\partial V_{\tau}}{\partial x^{\gamma}},$$

because the symbols used for dummy indices (σ and τ) are irrelevant. The fourth term of Eq. 6.23 equals the third term of Eq. 6.26:

$$-\Gamma^{\tau}_{\alpha\gamma}\frac{\partial V_{\tau}}{\partial x^{\beta}} = -\Gamma^{\sigma}_{\alpha\gamma}\frac{\partial V_{\sigma}}{\partial x^{\beta}},$$

for the same reason. The fifth terms are not equal:

$$\Gamma^{\tau}_{\alpha\gamma}\Gamma^{\sigma}_{\tau\beta}V_{\sigma} \neq \Gamma^{\tau}_{\alpha\beta}\Gamma^{\sigma}_{\tau\gamma}V_{\sigma}.$$

But the sixth terms are equal:

$$-\Gamma^{\eta}_{\beta\gamma}\frac{\partial V_{\alpha}}{\partial x^{\eta}} = -\Gamma^{\eta}_{\gamma\beta}\frac{\partial V_{\alpha}}{\partial x^{\eta}},$$

because Christoffel symbols are symmetric in their lower indices. The seventh terms are equal for the same reason:

$$\Gamma^{\eta}_{\beta\gamma}\Gamma^{\sigma}_{\alpha\eta}V_{\sigma} = \Gamma^{\eta}_{\gamma\beta}\Gamma^{\sigma}_{\alpha\eta}V_{\sigma}.$$

So when the commutator AB−BA is formed, most of the terms cancel out, but the second and fifth terms remain after subtraction. Those terms are

$$\begin{aligned} V_{\alpha\beta;\gamma} - V_{\alpha\gamma;\beta} &= -\frac{\partial \Gamma^{\sigma}_{\alpha\beta}}{\partial x^{\gamma}}V_{\sigma} + \frac{\partial \Gamma^{\sigma}_{\alpha\gamma}}{\partial x^{\beta}}V_{\sigma} + \Gamma^{\tau}_{\alpha\gamma}\Gamma^{\sigma}_{\tau\beta}V_{\sigma} - \Gamma^{\tau}_{\alpha\beta}\Gamma^{\sigma}_{\tau\gamma}V_{\sigma} \\ &= \left(\frac{\partial \Gamma^{\sigma}_{\alpha\gamma}}{\partial x^{\beta}} - \frac{\partial \Gamma^{\sigma}_{\alpha\beta}}{\partial x^{\gamma}} + \Gamma^{\tau}_{\alpha\gamma}\Gamma^{\sigma}_{\tau\beta} - \Gamma^{\tau}_{\alpha\beta}\Gamma^{\sigma}_{\tau\gamma}\right)V_{\sigma}. \end{aligned} \tag{6.27}$$

The terms within the parentheses define the Riemann curvature tensor:

$$R^{\sigma}{}_{\alpha\beta\gamma} \equiv \frac{\partial \Gamma^{\sigma}_{\alpha\gamma}}{\partial x^{\beta}} - \frac{\partial \Gamma^{\sigma}_{\alpha\beta}}{\partial x^{\gamma}} + \Gamma^{\tau}_{\alpha\gamma}\Gamma^{\sigma}_{\tau\beta} - \Gamma^{\tau}_{\alpha\beta}\Gamma^{\sigma}_{\tau\gamma}. \tag{6.28}$$

If you're wondering why the curvature tensor involves the derivative of Christoffel symbols, consider this: in any space, you can always define a coordinate system for which the Christoffel symbols are all zero at some point. But unless the space is flat, the Christoffel symbols will not be zero at all other locations, which means that the partial derivatives of the Christoffel symbols will not be zero. So a necessary and sufficient condition for flat space is that

$$R^{\sigma}_{\alpha\beta\gamma} = 0. \tag{6.29}$$

Another tensor related to the Riemann curvature tensor is the Ricci tensor, which you can find by contracting the Riemann tensor along the σ and β indices. In four dimensions, this is

$$R_{\alpha\gamma} \equiv R^{\sigma}_{\alpha\sigma\gamma} = R^{1}_{\alpha 1\gamma} + R^{2}_{\alpha 2\gamma} + R^{3}_{\alpha 3\gamma} + R^{4}_{\alpha 4\gamma}. \tag{6.30}$$

If you contract the Ricci tensor by raising one index and setting it equal to the other, the result is the Ricci scalar. Again in four dimensions, this is

$$R \equiv g^{\alpha\gamma} R_{\alpha\gamma} = R_\gamma^\gamma = R_1^1 + R_2^2 + R_3^3 + R_4^4. \tag{6.31}$$

Finally, the tensor known as the "Einstein tensor" can be written as a combination of the Ricci tensor, the Ricci scalar, and the metric:

$$G_{\alpha\gamma} \equiv R_{\alpha\gamma} - \frac{1}{2} R g_{\alpha\gamma}. \tag{6.32}$$

This is the tensor that appears in Einstein's field equation for General Relativity, often written as

$$G_{\mu\nu} + \Gamma g_{\mu\nu} = \frac{8\pi G}{c^4} T_{\mu\nu}, \tag{6.33}$$

where $T_{\mu\nu}$ is the energy-momentum tensor and Γ is the "cosmological constant" introduced by Einstein to maintain a static Universe. It is this equation that gives rise to the first half of the concise statement of General Relativity: "Matter tells spacetime how to curve, and spacetime tells matter how to move."

To appreciate the full content of the Riemann tensor, consider a two-dimensional space that is the surface of a sphere. The metric for such a space is

$$ds^2 = a^2 d\theta^2 + a^2 \sin^2(\theta) d\phi^2,$$

from which the components of the metric tensor may be found to be

$$\begin{aligned} g_{\theta\theta} &= a^2, \\ g_{\theta\phi} &= g_{\phi\theta} = 0, \\ g_{\phi\phi} &= a^2 \sin^2(\theta). \end{aligned} \tag{6.34}$$

Inserting these values into the equation for Christoffel symbols gives

$$\Gamma_{ij}^l = \frac{1}{2} g^{kl} \left[\frac{\partial g_{ik}}{\partial x^j} + \frac{\partial g_{jk}}{\partial x^i} - \frac{\partial g_{ij}}{\partial x^k} \right].$$

Even in two dimensions, writing out all the terms of the Christoffel symbols can be something of a chore:

$$\Gamma_{\theta\theta}^\theta = \frac{1}{2} \left[g^{\theta\theta} \frac{\partial g_{\theta\theta}}{\partial \theta} + g^{\phi\theta} \frac{\partial g_{\theta\phi}}{\partial \theta} + g^{\theta\theta} \frac{\partial g_{\theta\theta}}{\partial \theta} + g^{\phi\theta} \frac{\partial g_{\theta\phi}}{\partial \theta} - g^{\theta\theta} \frac{\partial g_{\theta\theta}}{\partial \theta} - g^{\phi\theta} \frac{\partial g_{\theta\theta}}{\partial \phi} \right],$$

$$\Gamma_{\theta\phi}^\theta = \frac{1}{2} \left[g^{\theta\theta} \frac{\partial g_{\theta\theta}}{\partial \phi} + g^{\phi\theta} \frac{\partial g_{\theta\phi}}{\partial \phi} + g^{\theta\theta} \frac{\partial g_{\phi\theta}}{\partial \theta} + g^{\phi\theta} \frac{\partial g_{\phi\phi}}{\partial \theta} - g^{\theta\theta} \frac{\partial g_{\theta\phi}}{\partial \theta} - g^{\phi\theta} \frac{\partial g_{\theta\phi}}{\partial \phi} \right],$$

$$\Gamma_{\phi\theta}^\theta = \frac{1}{2} \left[g^{\theta\theta} \frac{\partial g_{\phi\theta}}{\partial \theta} + g^{\phi\theta} \frac{\partial g_{\phi\phi}}{\partial \theta} + g^{\theta\theta} \frac{\partial g_{\theta\theta}}{\partial \phi} + g^{\phi\theta} \frac{\partial g_{\theta\phi}}{\partial \phi} - g^{\theta\theta} \frac{\partial g_{\phi\theta}}{\partial \theta} - g^{\phi\theta} \frac{\partial g_{\phi\theta}}{\partial \phi} \right],$$

$$\Gamma_{\theta\theta}^\phi = \frac{1}{2} \left[g^{\theta\phi} \frac{\partial g_{\theta\theta}}{\partial \theta} + g^{\phi\phi} \frac{\partial g_{\theta\phi}}{\partial \theta} + g^{\theta\phi} \frac{\partial g_{\theta\theta}}{\partial \theta} + g^{\phi\phi} \frac{\partial g_{\theta\phi}}{\partial \theta} - g^{\theta\phi} \frac{\partial g_{\theta\theta}}{\partial \theta} - g^{\phi\phi} \frac{\partial g_{\theta\theta}}{\partial \phi} \right],$$

$$\Gamma^{\phi}_{\theta\phi} = \frac{1}{2}\left[g^{\theta\phi}\frac{\partial g_{\theta\theta}}{\partial\phi} + g^{\phi\phi}\frac{\partial g_{\theta\phi}}{\partial\phi} + g^{\theta\phi}\frac{\partial g_{\phi\theta}}{\partial\theta} + g^{\phi\phi}\frac{\partial g_{\phi\phi}}{\partial\theta} - g^{\theta\phi}\frac{\partial g_{\theta\phi}}{\partial\theta} - g^{\phi\phi}\frac{\partial g_{\theta\phi}}{\partial\phi}\right],$$

$$\Gamma^{\phi}_{\phi\theta} = \frac{1}{2}\left[g^{\theta\phi}\frac{\partial g_{\phi\theta}}{\partial\theta} + g^{\phi\phi}\frac{\partial g_{\phi\phi}}{\partial\theta} + g^{\theta\phi}\frac{\partial g_{\theta\theta}}{\partial\phi} + g^{\phi\phi}\frac{\partial g_{\theta\phi}}{\partial\phi} - g^{\theta\phi}\frac{\partial g_{\phi\theta}}{\partial\theta} - g^{\phi\phi}\frac{\partial g_{\phi\theta}}{\partial\phi}\right],$$

$$\Gamma^{\theta}_{\phi\phi} = \frac{1}{2}\left[g^{\theta\theta}\frac{\partial g_{\phi\theta}}{\partial\phi} + g^{\phi\theta}\frac{\partial g_{\phi\phi}}{\partial\phi} + g^{\theta\theta}\frac{\partial g_{\phi\theta}}{\partial\phi} + g^{\phi\theta}\frac{\partial g_{\phi\phi}}{\partial\phi} - g^{\theta\theta}\frac{\partial g_{\phi\phi}}{\partial\theta} - g^{\phi\theta}\frac{\partial g_{\phi\phi}}{\partial\phi}\right],$$

$$\Gamma^{\phi}_{\phi\phi} = \frac{1}{2}\left[g^{\theta\phi}\frac{\partial g_{\phi\theta}}{\partial\phi} + g^{\phi\phi}\frac{\partial g_{\phi\phi}}{\partial\phi} + g^{\theta\phi}\frac{\partial g_{\phi\theta}}{\partial\phi} + g^{\phi\phi}\frac{\partial g_{\phi\phi}}{\partial\phi} - g^{\theta\phi}\frac{\partial g_{\phi\phi}}{\partial\theta} - g^{\phi\phi}\frac{\partial g_{\phi\phi}}{\partial\phi}\right].$$

But given the metric tensor components shown in Eq. 6.34, all the partial derivatives except those involving $\frac{\partial g_{\phi\phi}}{\partial\theta}$ are zero, as are any terms involving $g_{\theta\phi}$ or $g_{\phi\theta}$. That leaves only three non-zero Christoffel symbols, which are

$$\begin{aligned}
\Gamma^{\phi}_{\theta\phi} &= \left(\frac{1}{2}\right) g^{\phi\phi}\frac{\partial g_{\phi\phi}}{\partial\theta} \\
&= \left(\frac{1}{2}\right)\frac{1}{a^2\sin^2(\theta)}[2a^2\sin(\theta)\cos(\theta)] = \frac{\cos(\theta)}{\sin(\theta)} = \cot(\theta), \\
\Gamma^{\phi}_{\phi\theta} &= \left(\frac{1}{2}\right) g^{\phi\phi}\frac{\partial g_{\phi\phi}}{\partial\theta} \\
&= \cot(\theta), \\
\Gamma^{\theta}_{\phi\phi} &= \left(\frac{1}{2}\right) - g^{\theta\theta}\frac{\partial g_{\phi\phi}}{\partial\theta} \\
&= -\left(\frac{1}{2}\right)\frac{1}{a^2}[2a^2\sin(\theta)\cos(\theta)] = -\sin(\theta)\cos(\theta).
\end{aligned}$$

With the Christoffel symbols for the spherical surface in hand, the components of the Riemann curvature tensor may be found using

$$R^{\sigma}{}_{\alpha\beta\gamma} \equiv \frac{\partial\Gamma^{\sigma}_{\alpha\gamma}}{\partial x^{\beta}} - \frac{\partial\Gamma^{\sigma}_{\alpha\beta}}{\partial x^{\gamma}} + \Gamma^{\tau}_{\alpha\gamma}\Gamma^{\sigma}_{\tau\beta} - \Gamma^{\tau}_{\alpha\beta}\Gamma^{\sigma}_{\tau\gamma}.$$

As in most tensor equations, the full content of this tensor can only be appreciated by writing out the components. Not only must you allow each of the indices σ, α, β, and γ to represent both θ and ϕ, you must also allow the dummy index τ to represent both θ and ϕ and then sum those terms. Hence in two-dimensional space, the last two terms of the Riemann tensor equation (those involving the products of the Christoffel symbols) become four terms, making a total of six terms for each set of indices. The first eight components of the Riemann tensor can be found by setting σ equal to θ and letting the other indices represent both θ and ϕ:

$$R^{\theta}{}_{\theta\theta\theta} = \frac{\partial\Gamma^{\theta}_{\theta\theta}}{\partial\theta} - \frac{\partial\Gamma^{\theta}_{\theta\theta}}{\partial\theta} + \Gamma^{\theta}_{\theta\theta}\Gamma^{\theta}_{\theta\theta} + \Gamma^{\phi}_{\theta\theta}\Gamma^{\theta}_{\phi\theta} - \Gamma^{\theta}_{\theta\theta}\Gamma^{\theta}_{\theta\theta} - \Gamma^{\phi}_{\theta\theta}\Gamma^{\theta}_{\phi\theta},$$

$$R^{\theta}{}_{\theta\theta\phi} = \frac{\partial\Gamma^{\theta}_{\theta\phi}}{\partial\theta} - \frac{\partial\Gamma^{\theta}_{\theta\theta}}{\partial\phi} + \Gamma^{\theta}_{\theta\phi}\Gamma^{\theta}_{\theta\theta} + \Gamma^{\phi}_{\theta\phi}\Gamma^{\theta}_{\phi\theta} - \Gamma^{\theta}_{\theta\theta}\Gamma^{\theta}_{\theta\phi} - \Gamma^{\phi}_{\theta\theta}\Gamma^{\theta}_{\phi\phi},$$

$$R^{\theta}{}_{\theta\phi\theta} = \frac{\partial \Gamma^{\theta}_{\theta\theta}}{\partial \phi} - \frac{\partial \Gamma^{\theta}_{\theta\phi}}{\partial \theta} + \Gamma^{\theta}_{\theta\theta}\Gamma^{\theta}_{\theta\phi} + \Gamma^{\phi}_{\theta\theta}\Gamma^{\theta}_{\phi\phi} - \Gamma^{\theta}_{\theta\phi}\Gamma^{\theta}_{\theta\theta} - \Gamma^{\phi}_{\theta\phi}\Gamma^{\theta}_{\phi\theta},$$

$$R^{\theta}{}_{\phi\theta\theta} = \frac{\partial \Gamma^{\theta}_{\phi\theta}}{\partial \theta} - \frac{\partial \Gamma^{\theta}_{\phi\theta}}{\partial \phi} + \Gamma^{\theta}_{\phi\theta}\Gamma^{\theta}_{\theta\theta} + \Gamma^{\phi}_{\phi\theta}\Gamma^{\theta}_{\phi\theta} - \Gamma^{\theta}_{\phi\theta}\Gamma^{\theta}_{\theta\theta} - \Gamma^{\phi}_{\phi\theta}\Gamma^{\theta}_{\phi\theta},$$

$$R^{\theta}{}_{\theta\phi\phi} = \frac{\partial \Gamma^{\theta}_{\theta\phi}}{\partial \phi} - \frac{\partial \Gamma^{\theta}_{\theta\phi}}{\partial \phi} + \Gamma^{\theta}_{\theta\phi}\Gamma^{\theta}_{\theta\phi} + \Gamma^{\phi}_{\theta\phi}\Gamma^{\theta}_{\phi\phi} - \Gamma^{\theta}_{\theta\phi}\Gamma^{\theta}_{\theta\phi} - \Gamma^{\phi}_{\theta\phi}\Gamma^{\theta}_{\phi\phi},$$

$$R^{\theta}{}_{\phi\theta\phi} = \frac{\partial \Gamma^{\theta}_{\phi\phi}}{\partial \theta} - \frac{\partial \Gamma^{\theta}_{\phi\theta}}{\partial \phi} + \Gamma^{\theta}_{\phi\phi}\Gamma^{\theta}_{\theta\theta} + \Gamma^{\phi}_{\phi\phi}\Gamma^{\theta}_{\phi\theta} - \Gamma^{\theta}_{\phi\theta}\Gamma^{\theta}_{\theta\phi} - \Gamma^{\phi}_{\phi\theta}\Gamma^{\theta}_{\phi\phi},$$

$$R^{\theta}{}_{\phi\phi\theta} = \frac{\partial \Gamma^{\theta}_{\phi\theta}}{\partial \phi} - \frac{\partial \Gamma^{\theta}_{\phi\phi}}{\partial \theta} + \Gamma^{\theta}_{\phi\theta}\Gamma^{\theta}_{\theta\phi} + \Gamma^{\phi}_{\phi\theta}\Gamma^{\theta}_{\phi\phi} - \Gamma^{\theta}_{\phi\phi}\Gamma^{\theta}_{\theta\theta} - \Gamma^{\phi}_{\phi\phi}\Gamma^{\theta}_{\phi\theta},$$

$$R^{\theta}{}_{\phi\phi\phi} = \frac{\partial \Gamma^{\theta}_{\phi\phi}}{\partial \phi} - \frac{\partial \Gamma^{\theta}_{\phi\phi}}{\partial \phi} + \Gamma^{\theta}_{\phi\phi}\Gamma^{\theta}_{\theta\phi} + \Gamma^{\phi}_{\phi\phi}\Gamma^{\theta}_{\phi\phi} - \Gamma^{\theta}_{\phi\phi}\Gamma^{\theta}_{\theta\phi} - \Gamma^{\phi}_{\phi\phi}\Gamma^{\theta}_{\phi\phi}.$$

Inserting the Christoffel symbols found above, you can see that the non-zero components are

$$R^{\theta}{}_{\phi\theta\phi} = \frac{\partial \Gamma^{\theta}_{\phi\phi}}{\partial \theta} - \Gamma^{\phi}_{\phi\theta}\Gamma^{\theta}_{\phi\phi},$$

$$R^{\theta}{}_{\phi\phi\theta} = -\frac{\partial \Gamma^{\theta}_{\phi\phi}}{\partial \theta} + \Gamma^{\phi}_{\phi\theta}\Gamma^{\theta}_{\phi\phi}.$$

And since

$$\frac{\partial \Gamma^{\theta}_{\phi\phi}}{\partial \theta} = \sin^2(\theta) - \cos^2(\theta),$$

and

$$\Gamma^{\phi}_{\phi\theta}\Gamma^{\theta}_{\phi\phi} = -\cos^2(\theta),$$

this means the surviving terms from the $\sigma = \theta$ group are

$$R^{\theta}{}_{\phi\theta\phi} = [\sin^2(\theta) - \cos^2(\theta)] - [-\cos^2(\theta)] = \sin^2(\theta),$$

$$R^{\theta}{}_{\phi\phi\theta} = -[\sin^2(\theta) - \cos^2(\theta)] + [-\cos^2(\theta)] = -\sin^2(\theta).$$

Now allowing σ to equal ϕ, the other eight terms are

$$R^{\phi}{}_{\theta\theta\theta} = \frac{\partial \Gamma^{\phi}_{\theta\theta}}{\partial \theta} - \frac{\partial \Gamma^{\phi}_{\theta\theta}}{\partial \theta} + \Gamma^{\theta}_{\theta\theta}\Gamma^{\phi}_{\theta\theta} + \Gamma^{\phi}_{\theta\theta}\Gamma^{\phi}_{\phi\theta} - \Gamma^{\theta}_{\theta\theta}\Gamma^{\phi}_{\theta\theta} - \Gamma^{\phi}_{\theta\theta}\Gamma^{\phi}_{\phi\theta},$$

$$R^{\phi}{}_{\theta\theta\phi} = \frac{\partial \Gamma^{\phi}_{\theta\phi}}{\partial \theta} - \frac{\partial \Gamma^{\phi}_{\theta\theta}}{\partial \phi} + \Gamma^{\theta}_{\theta\phi}\Gamma^{\phi}_{\theta\theta} + \Gamma^{\phi}_{\theta\phi}\Gamma^{\phi}_{\phi\theta} - \Gamma^{\theta}_{\theta\theta}\Gamma^{\phi}_{\theta\phi} - \Gamma^{\phi}_{\theta\theta}\Gamma^{\phi}_{\phi\phi},$$

$$R^{\phi}{}_{\theta\phi\theta} = \frac{\partial \Gamma^{\phi}_{\theta\theta}}{\partial \phi} - \frac{\partial \Gamma^{\phi}_{\theta\phi}}{\partial \theta} + \Gamma^{\theta}_{\theta\theta}\Gamma^{\phi}_{\theta\phi} + \Gamma^{\phi}_{\theta\theta}\Gamma^{\phi}_{\phi\phi} - \Gamma^{\theta}_{\theta\phi}\Gamma^{\phi}_{\theta\theta} - \Gamma^{\phi}_{\theta\phi}\Gamma^{\phi}_{\phi\theta},$$

$$R^{\phi}{}_{\phi\theta\theta} = \frac{\partial \Gamma^{\phi}_{\phi\theta}}{\partial\theta} - \frac{\partial \Gamma^{\phi}_{\phi\theta}}{\partial\theta} + \Gamma^{\theta}_{\phi\theta}\Gamma^{\phi}_{\theta\theta} + \Gamma^{\phi}_{\phi\theta}\Gamma^{\phi}_{\phi\theta} - \Gamma^{\theta}_{\phi\theta}\Gamma^{\phi}_{\theta\theta} - \Gamma^{\phi}_{\phi\theta}\Gamma^{\phi}_{\phi\theta},$$

$$R^{\phi}{}_{\theta\phi\phi} = \frac{\partial \Gamma^{\phi}_{\theta\phi}}{\partial\phi} - \frac{\partial \Gamma^{\phi}_{\theta\phi}}{\partial\phi} + \Gamma^{\theta}_{\theta\phi}\Gamma^{\phi}_{\theta\phi} + \Gamma^{\phi}_{\theta\phi}\Gamma^{\phi}_{\phi\phi} - \Gamma^{\theta}_{\phi\theta}\Gamma^{\phi}_{\theta\phi} - \Gamma^{\phi}_{\theta\phi}\Gamma^{\phi}_{\phi\phi},$$

$$R^{\phi}{}_{\phi\theta\phi} = \frac{\partial \Gamma^{\phi}_{\phi\phi}}{\partial\theta} - \frac{\partial \Gamma^{\phi}_{\phi\theta}}{\partial\phi} + \Gamma^{\theta}_{\phi\phi}\Gamma^{\phi}_{\theta\theta} + \Gamma^{\phi}_{\phi\phi}\Gamma^{\phi}_{\phi\theta} - \Gamma^{\theta}_{\phi\theta}\Gamma^{\phi}_{\theta\phi} - \Gamma^{\phi}_{\phi\theta}\Gamma^{\phi}_{\phi\phi},$$

$$R^{\phi}{}_{\phi\phi\theta} = \frac{\partial \Gamma^{\phi}_{\phi\theta}}{\partial\phi} - \frac{\partial \Gamma^{\phi}_{\phi\phi}}{\partial\theta} + \Gamma^{\theta}_{\phi\theta}\Gamma^{\phi}_{\theta\phi} + \Gamma^{\phi}_{\phi\theta}\Gamma^{\phi}_{\phi\phi} - \Gamma^{\theta}_{\phi\phi}\Gamma^{\phi}_{\theta\theta} - \Gamma^{\phi}_{\phi\phi}\Gamma^{\phi}_{\phi\theta},$$

$$R^{\phi}{}_{\phi\phi\phi} = \frac{\partial \Gamma^{\phi}_{\phi\phi}}{\partial\phi} - \frac{\partial \Gamma^{\phi}_{\phi\phi}}{\partial\phi} + \Gamma^{\theta}_{\phi\phi}\Gamma^{\phi}_{\theta\phi} + \Gamma^{\phi}_{\phi\phi}\Gamma^{\phi}_{\phi\phi} - \Gamma^{\theta}_{\phi\phi}\Gamma^{\phi}_{\theta\phi} - \Gamma^{\phi}_{\phi\phi}\Gamma^{\phi}_{\phi\phi}.$$

Again inserting the Christoffel symbols, the non-zero terms are found to be

$$R^{\phi}{}_{\theta\theta\phi} = \frac{\partial \Gamma^{\phi}_{\theta\phi}}{\partial\theta} + \Gamma^{\phi}_{\theta\phi}\Gamma^{\phi}_{\phi\theta},$$

$$R^{\phi}{}_{\theta\phi\theta} = -\frac{\partial \Gamma^{\phi}_{\theta\phi}}{\partial\theta} - \Gamma^{\phi}_{\theta\phi}\Gamma^{\phi}_{\phi\theta}.$$

And since

$$\frac{\partial \Gamma^{\phi}_{\theta\phi}}{\partial\theta} = -\frac{\sin(\theta)}{\sin(\theta)} - \frac{\cos^2(\theta)}{\sin^2(\theta)} = -[1+\cot^2(\theta)],$$

and

$$\Gamma^{\phi}_{\theta\phi}\Gamma^{\phi}_{\phi\theta} = \cot^2(\theta),$$

the surviving terms are

$$R^{\phi}{}_{\theta\theta\phi} = -[1+\cot^2(\theta)] + \cot^2(\theta) = -1,$$

$$R^{\phi}{}_{\theta\phi\theta} = [1+\cot^2(\theta)] - \cot^2(\theta) = 1.$$

As expected, a two-dimensional space with the metric of a sphere ($ds^2 = a^2 d\theta^2 + a^2 \sin^2(\theta) d\phi^2$) has non-zero components of the Riemann curvature tensor, confirming that this space is non-Euclidean.

You can see how to use these results to find the Ricci tensor and the Ricci scalar in the on-line solutions to the problems at the end of this chapter.

6.4 Chapter 6 problems

6.1 Find the inertia tensor for a cubical arrangement of eight identical masses with the origin of coordinates at one of the corners and the coordinate axes along the edges of the cube.

6.2 How would the moment of inertia tensor of Problem 6.1 change if one of the eight masses is removed?

6.3 Find the moment of inertia tensor for the arrangement of masses of Problem 6.2 if the coordinate system is rotated by 20 degrees about one of the coordinate axes (do this by finding the locations of the masses in the rotated coordinate system).

6.4 Use the similarity-transform approach to verify the moment of inertia tensor you found in Problem 6.3.

6.5 Show how the vector wave equation results from taking the curl of both sides of Faraday's Law and inserting the curl of the magnetic field from the Ampere–Maxwell Law.

6.6 If an observer in one coordinate system measures an electric field of 5 volts per meter in the z-direction and zero magnetic field, what electric and magnetic fields would be measured by a second observer moving at 1/4 the speed of light along the x-axis?

6.7 If an observer in one coordinate system measures a magnetic field of 1.5 tesla in the z-direction and zero electric field, what electric and magnetic fields would be measured by a second observer moving at 1/4 the speed of light along the x-axis?

6.8 Show that $\vec{E} \circ \vec{B}$ is invariant under Lorentz transformation.

6.9 The differential line element in 2-D Euclidean space may be expressed in polar coordinates as $ds^2 = dr^2 + r^2 d\theta^2$. Show that the Riemann curvature tensor equals zero in this case, as it must for any flat space.

6.10 Find the Ricci tensor and scalar for the 2-sphere of Section 6.3.

Further reading

Arfken, G. and Weber, H., *Mathematical Methods for Physicists*, Elsevier Academic Press 2005.
Boas, M., *Mathematical Methods in the Physical Sciences*, John Wiley and Sons 2006.
Borisenko, A. and Tarapov, I., *Vector and Tensor Analysis*, Dover Press 1979.
Carroll, S., *Spacetime and Geometry: An Introduction to General Relativity*, Benjamin-Cummings 2003.
Einstein, A., *The Meaning of Relativity*, Princeton University Press 2004.
Griffiths, D., *Introduction to Electrodynamics*, Benjamin-Cummings 1999.
Jackson, J., *Classical Electrodynamics*, John Wiley and Sons 1999.
Lieber, L., *The Einstein Theory of Relativity*, Paul Dry Books 2008.
Matthews, P., *Vector Calculus*, Springer-Verlag 1998.
McMahon, D., *Relativity Demystified*, McGraw-Hill 2006.
Morse, P. and Feshbach, H., *Methods of Theoretical Physics*, McGraw-Hill 1953.
Schutz, B., *A First Course in General Relativity*, Cambridge University Press 2009.
Spiegel, M., *Vector Analysis*, McGraw-Hill 1959.
Stroud, K., *Vector Analysis*, Industrial Press 2005.

Index

acceleration, 72
acceleration of gravity, 67
active transformation, 108
Ampere–Maxwell Law, 173
angular momentum, 160
angular velocity, 160
arithmetized space, 140
asymmetric top, 166

BAC minus CAB rule, 33, 162
basis vectors, 20
  as partial derivatives, 23
  dual, 113
  orthonormal, 21
basis-vector transformation, 105
bound vectors, 2

Cartesian coordinates
  unit vectors, 5
center of mass, 64
centrifugal force, 76
centripetal acceleration, 75
centripetal force, 76
chain rule, 41, 42
Christoffel, Elwin Bruno, 184
Christoffel symbols, 148
column vectors, 4
commutator, 186
continuity equation, 173
convergence, 46
coordinate-system transformation, 97
cosmological constant, 189
Coulomb constant, 85
covariant differentiation, 153, 186
  notation, 155
covectors, 156
Cramer's Rule, 118, 120
cross product, 27
curl, 50
curvilinear motion, 72
cylindrical coordinates, 17
  unit vectors, 19

del cross, 51
del dot, 48
del operator, 43
differential length element, 140
direct transformation, 108
direction cosines, 102
directional derivative
  as tangent vector, 43
directional derivatives, 41
divergence, 46
  of gradient, 54
dot product, 25
dual basis vectors, 113
dual contravariant electromagnetic field tensor, 178

Einstein, Albert, 123, 174, 183
Einstein summation convention, 123
Einstein tensor, 189
electric field, 81
electric force, 83
electric potential, 88
electromagnetic field tensor, 171, 178
electromagnetic wave equation, 174
electrostatic field, 83
equipotential surfaces, 88
Euclidean space, 185

Faraday's Law, 173
field
  definition, 81
  electric, 81
  electrostatic, 46, 83
  irrotational, 52, 88
  magnetic, 89
  magnetostatic, 89
  scalar, 44
  vector, 44
field lines, 83
four-current, 177
four-dimensional spacetime, 174
four-vector, 176
free vectors, 2
free-body diagram, 63
friction, 69
frictional force, 70

Galilean transformation, 175
Gauss's Law
  for electric fields, 87, 172
  for magnetic fields, 90, 172
General Relativity, 183
gradient, 44

inclined plane, 62
index notation, 122
index raising and lowering, 147
inertia tensor, 159, 164
inertial reference frame, 184
inner product, 138
inverse transformation, 105
irrotational fields, 52, 88

kinetic friction, 70
Kronecker Delta function, 139

Laplace, Pierre-Simon, 54
Laplace's Equation, 89
Laplacian, 54
  as difference from surrounding points, 57
  as divergence of gradient, 54
  as peak finder, 57
length contraction, 177
linearly independent vectors, 21
Lorentz, Hendrik, 176
Lorentz transform, 176
Lorentz transformation matrix, 181

magnetic field, 89
magnetic force, 91
magnetostatic field, 89
manifest covariance, 177
Maxwell, James Clerk, 46, 81, 171
Maxwell's Equations, 172
  tensor form, 178
metric tensor, 140
  notation, 140
Minkowski metric, 177
moment of inertia
  for a single particle, 160
moments of inertia, 164

nabla, 43
Newton, Isaac, 67
Newton's Second Law, 63
non-Cartesian coordinate systems
  cylindrical coordinates, 17
  polar coordinates, 15
  spherical coordinates, 19
non-Cartesian coordinates
  unit vectors, 14
non-orthogonal coordinate systems, 110
normal force, 63

one-forms, 156
operator, 44
operator equation, 43
ordinary derivatives, 35
orthogonal transformation, 110
outer product, 137

parallel projection, 111
parallel transport, 154, 185
parameterized curve, 42
partial derivatives, 35
  as basis vectors, 23
  as slope, 37
  chain rule, 41
  higher-order, 40
  mixed, 40
  notation, 35
passive transformation, 105
perpendicular projection, 112
Poisson's Equation, 88
polar coordinates, 15
  unit vectors, 16
principal axes, 166
principal moments, 166
principle of equivalence, 184
products of inertia, 164

Pythagorean theorem, 10

radial acceleration, 72
reciprocal basis vectors, 114
relativity of simultaneity, 177
Ricci scalar, 189
Ricci tensor, 188
Riemann, Bernhard, 184
Riemann curvature tensor, 183
right-hand rule, 28
rotor, 166
row vectors, 4

scalar, 4
  definition, 4, 133
  field, 44
  Ricci, 189
scalar product, 25
scalar triple product, 30
scale factors, 146
similarity transform, 170
sliding vectors, 3
space–time interval, 176
Special Relativity, 174
spherical coordinates, 19
  unit vectors, 19
spherical top, 166
static friction, 70
summation convention, 123
symmetric top, 166

tangential acceleration, 72
tensor, 4
  addition and subtraction, 135
  definition, 5, 134
  derivatives, 148
  Einstein, 189
  electromagnetic field, 171, 178
  higher-rank, 132
  inertia, 159, 164
  inner product, 138
  metric, 140
  multiplication, 137
  notation, 134
  rank, 5
  Ricci, 188
  Riemann curvature, 183
test charge, 82
time dilation, 177
top, 166
transformation
  basis-vector, 105
  coordinate-system, 97
  direct or active, 108
  equation, 102
  inverse or passive, 105
  matrix, 102
  orthogonal, 110
triple scalar product, 30, 116
triple vector product, 32

unit vectors
  Cartesian, 5
  non-Cartesian, 14

vector, 1
  addition, 11
    graphical, 12
    using components, 13
  as an ordered set, 3
  as derivative, 41
  basis, 2
  bound, 2
  column, 4
  components, 4, 7
    covariant and contravariant, 97, 105
  definition, 1, 133
  field, 44
  free, 2
  graphical depiction, 1
  linearly independent, 21
  multiplication by a scalar, 11
  notation, 1
  outer product, 137
  row, 4
  sliding, 3
  unit
    Cartesian, 5
    non-Cartesian, 14
vector components, 4, 7
  as projections onto coordinate axes, 8
vector field, 3
versors, 6

weighted linear combination, 101
work, 25

Printed in the United States
By Bookmasters